Mechanisms of Epithelial Defense

Chemical Immunology and Allergy

Vol. 86

KARGER

Mechanisms of Epithelial Defense

Volume Editors

Dieter Kabelitz Kiel
Jens-Michael Schröder Kiel

21 figures, 2 in color, and 4 tables, 2005

KARGER

Basel · Freiburg · Paris · London · New York ·
Bangalore · Bangkok · Singapore · Tokyo · Sydney

Chemical Immunology and Allergy
Formerly published as 'Progress in Allergy' (Founded 1939)
continued 1990–2002 as 'Chemical Immunology'
Edited by Paul Kalos 1939–1988, Byron H. Waksman 1962–2000

Dieter Kabelitz
Institute of Immunology
UKSH Campus Kiel
Michaelisstrasse 5
Kiel (Germany)

Jens-Michael Schröder
Clinical Research Unit at the Department of Dermatology
University Hospital Schleswig-Holstein
Campus Kiel, Schittenhelmstrasse 7
Kiel (Germany)

Bibliographic Indices. This publication is listed in bibliographic services, including Current Contents® and Index Medicus.

Drug Dosage. The authors and the publisher have exerted every effort to ensure that drug selection and dosage set forth in this text are in accord with current recommendations and practice at the time of publication. However, in view of ongoing research, changes in government regulations, and the constant flow of information relating to drug therapy and drug reactions, the reader is urged to check the package insert for each drug for any change in indications and dosage and for added warnings and precautions. This is particularly important when the recommended agent is a new and/or infrequently employed drug.

www.karger.com
Printed in Switzerland on acid-free paper by Reinhardt Druck, Basel
ISSN 1660–2242
ISBN 3–8055–7862–8

Contents

Preface

Epithelia constitute the body's contact zone with a hostile environment, full of pathogenic microbes ready to attack the body. While it has been known for a long time that various physicochemical features including low pH contribute to the naturally protective milieu of the skin, it is a rather recent discovery that skin and other epithelia harbor a large array of naturally occurring antimicrobial peptides. In recent years, various classes of constitutive and inducible antimicrobial peptides have been identified, and the emerging scenario indicates that such peptides are instrumental in protecting the organism against infection and maintaining the homeostasis. Thus, it is very clear now that antimicrobial peptides are in the center of epithelial defense mechanisms and, more generally speaking, form an important part of the innate immune system. Another very important discovery of recent years which has revolutionized our appreciation of the innate immune system is the identification of a set of pattern recognition receptors which display specificity for defined microbial molecules including bacterial lipopolysaccharide, bacterial DNA, viral RNA, and others. These Toll-like receptors and intracellular recognition receptors such as NODs are early sensors of microbial material in both, epithelial cells and dendritic cells, and thus initiate innate immune responses way before the specific or adaptive immune system is alerted. Interestingly, and in line with their essential functions, both antimicrobial peptides and microbe-sensing Toll-like receptors have been conserved throughout evolution. In addition to these basic and ancient epithelial defense mechanisms, it has been appreciated that subsets of lymphocytes also contribute to local immune surveillance and defense in epithelia. While the mouse skin harbors a specialized population of γδ T lymphocytes termed dendritic epidermal T cells (DETC), subsets of γδ T cells expressing a T cell receptor different from those of peripheral blood γδ T cells also localize to the human intestinal epithelia. Various arguments support the idea that such γδ T cells form a bridge between the innate and adaptive immune system and, in addition, play a crucial role in local immune surveillance and in maintaining the epithelial integrity.

It is now realized that these multiple and at first glance independent mechanisms effectively protect the epithelia from infection and contribute to tissue repair. As a consequence, these defense systems are also involved in pathophysiology, and we are only beginning to understand their respective significance in disease. This book contains a collection of up-to-date reviews dealing with the roles of antimicrobial peptides, Toll-like receptors, and γδ T cells in epithelial defense and infection. We believe that the focus on these three mechanistic principles in the context of developmental and disease-related aspects offers interesting insights into the epithelial defense system in physiological and pathophysiological conditions.

Dieter Kabelitz
Jens-Michael Schröder
Kiel

Kabelitz D, Schröder J-M (eds): Mechanisms of Epithelial Defense.
Chem Immunol Allergy. Basel, Karger, 2005, vol 86, pp 1–21

Antimicrobial Peptides in *Drosophila*: Structures, Activities and Gene Regulation

Jean-Luc Imler[a], *Philippe Bulet*[b]

[a]CNRS UPR9022, Institut de Biologie Moléculaire et Cellulaire, Strasbourg, France;
[b]Atheris Laboratories, Bernex, Geneva, Switzerland

Abstract

The production of antimicrobial peptides (AMPs) is an important aspect of host-defence in multicellular organisms. Biochemical analysis of the hemolymph of the fruit-fly *Drosophila melanogaster* and other Diptera has led to the discovery of eight classes of AMPs. These peptides can be grouped into three families based on their main biological targets, gram-positive bacteria (defensin), gram-negative bacteria (cecropins, drosocin, attacins, diptericin, MPAC), or fungi (drosomycin, metchnikowin). *Drosophila* AMPs are synthesized by the fat body in response to infection, and secreted into the blood. Most of them can also be induced in surface epithelia in a tissue-specific manner. Finally, some of them are constitutively expressed in defined tissues, such as the salivary glands or the reproductive tract. We review here the structures and activities of these AMPs, as well as the signalling cascades, which lead to their induction upon detection of infectious non-self.

Introduction

Innate immunity is an ancient metazoan feature, which allows animals to recognize typical chemical structures presented by invading micro-organisms, and to activate a sophisticated set of responses to control the infection. In vertebrates, these responses include the induction of cytokines and co-stimulatory molecules that will trigger adaptive immunity. The fruit-fly *Drosophila melanogaster* has emerged in recent years as an original and powerful model to study the evolutionary conserved genetic and molecular mechanisms operating in innate immunity. Higher insects control infection by an array of innate immune reactions that include: (1) phagocytosis and encapsulation by blood cells; (2) proteolytic cascades leading to coagulation and melanisation, and (3) secretion of a cocktail of potent AMPs [1].

AMPs are innate host defence molecules that are effective on bacteria (gram-positive, gram-negative), fungi (yeasts and filamentous) and parasites, and in some cases on enveloped viruses. They are found in evolutionarily diverse organisms ranging from prokaryotes to invertebrates and vertebrates, and to plants [2–5]. AMPs are expressed in many types of cells and secretions. Humans express several families of AMPs in myeloid cells. Over the past several years, we have come to realize that various epithelial surfaces from invertebrates and vertebrates can also express their own battery of defensive molecules. In humans, the three AMP families are (1) defensins (six α-defensins and more than 20β-defensins were identified in the human genome) characterized by a β-sheet conformation stabilized by a six cysteine/three-disulfide pattern [5]; (2) cathelicidins (hCAp-18/LL-37 from human neutrophils) characterized by a conserved precursor motif named cathelin domain [5], and (3) histatins from saliva (three major forms among 12 variants) that are histidine-rich peptides. In terms of structural diversity, the human arsenal in AMPs is rather limited (three main classes) compared to that of the fruit-fly *D. melanogaster*. To date, eight distinct classes of AMPs have been identified in *D. melanogaster* [1, 6], which can be classified in three groups depending on their main microbial targets. *Drosophila* defensin is active against gram-positive bacteria, while drosocin, cecropins, attacins, diptericins and MPAC (truncated post-translationally modified pro-domain of attacin C) are active against gram-negative bacteria, and drosomycin and metchnikowin efficient against fungi (fig. 1). We review below the structure and activity of *Drosophila* AMPs, before turning to the molecular and genetic mechanisms that trigger their expression in response to infection.

Drosophila Defensin, An Anti-Gram-Positive Peptide

The *Drosophila* defensin is a 40-residue peptide with three internal disulfide bridges (Cys^1-Cys^4, Cys^2-Cys^5, Cys^3-Cys^6). Its three-dimensional structure has not been established but by homology to the *Protophormia terranovae* defensin A, *Drosophila* defensin may consist of an α-helical domain linked to antiparallel β-strands by two disulfide bridges (fig. 1) giving the motif termed cysteine-stabilized α-helix/β-sheet (CSαβ). This peculiar scaffold, present in all invertebrate defensins [7] as well as in plant defensins [8], is not observed in vertebrate defensins [5]. The global fold for vertebrate α- and β-defensins includes a well-defined three-stranded antiparallel β-sheet and for some of them an additional short N-terminal domain with an α-helical propensity.

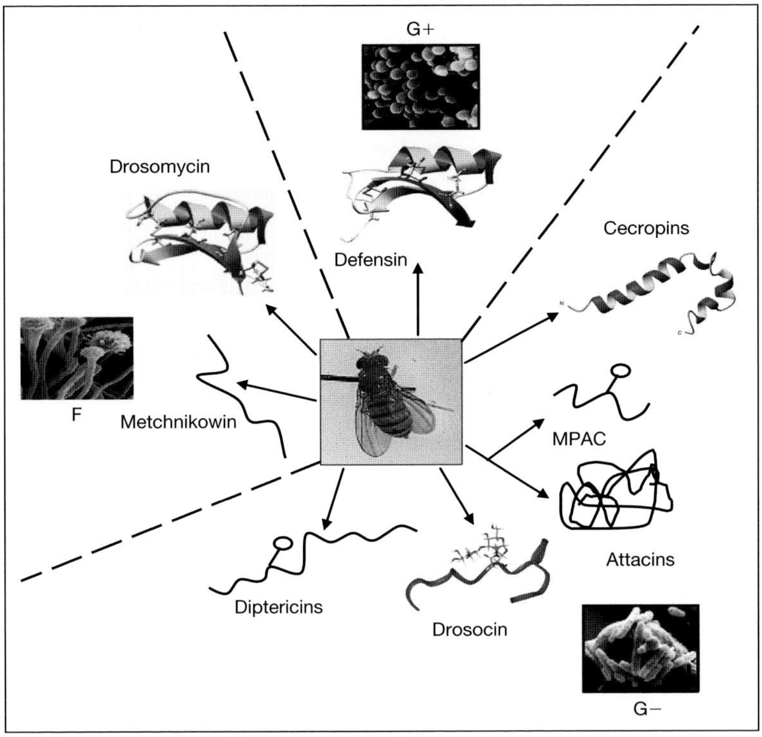

Fig. 1. The different families of AMPs produced by the fat body of *Drosophila melanogaster* in response to an experimental infection. Drosomycin and metchnikowin have antifungal properties (F), while defensin is active preferentially against gram-positive bacteria (G+). O-glycosylated drosocin, cecropins, MPAC (maturated pro-domain of attacin C), and by homology diptericins and attacins are affecting a large panel of gram-negative strains (G−) and a limited number of gram-positive bacteria. O-glycosylation of diptericins and MPAC is indicated. The branched arrow means that MPAC and attacins are originated from a same precursor.

D. melanogaster defensin is identical to the *D. simulans* one and has more than 70% homology with *Stomoxys* defensins, *Sarcophaga peregrina* defensin/sapecin A and *Protophormia* defensins. Important homologies (over 60%) also exist between the fruit-fly defensin and the defensins from *Nematocera*. Insect defensins are active on a large panel of gram-positive bacterial strains, and on a limited number of gram-negative strains and filamentous fungi. To date, no activity has been reported on yeast cells. Below physiological salt concentrations, insect defensins have, within a minute of contact with the pathogenic organism, a lytic effect at a micromolar level (0.1–10 μM). By homology to *Protophormia* defensin, *Drosophila* defensin may disrupt the permeability barrier of the cytoplasmic membrane of bacteria, resulting in a loss of cytoplasmic potassium, a partial depolarization of the inner membrane, a decrease in cytoplasmic ATP, and an inhibition of respiration [8]. Thus, increasing the salt at a relevant physiological concentration dramatically reduces the efficacy of these molecules. Such an observation is not limited to invertebrate defensins as this is also reported for vertebrate and plant defensins [9]. Mature *Drosophila* defensin that has an hemolymph titer of 1 μM, was partially characterized from an acidic extract of experimentally infected flies and in vitro antimicrobial assays, and its full primary sequence totally elucidated from cDNA cloning. *Drosophila* defensin is synthesized as a 92-residue precursor containing an N-terminal signal-peptide (20 residues) followed by a pro-domain (32 residues) and the mature defensin (fig. 2). The precursor polypeptide is encoded by an intronless gene (*Def*) mapping on chromosome 2R at position 46D9.

The Antifungal Drosomycin

Drosomycin is a 44-residue strictly antifungal peptide with a CSαβ motif as observed in insect defensins but it has the peculiarity to contain an additional disulfide bridge locking the cysteine residue at position two (Cys^1) with the C-terminal cysteine residue Cys^8. This gives the following disulfide pairing Cys^1-Cys^8, Cys^2-Cys^5, Cys^3-Cys^6 and Cys^4-Cys^7, which is identical to the one observed in plant defensins. Nuclear magnetic resonance (NMR) spectroscopy studies and molecular modelling on recombinant drosomycin reveals a βαββ scaffold. Compared to insect defensins, an additional short β-strand is present at the N-terminus of the molecule (fig. 1). Apart from the cysteine residues, drosomycin has no sequence similarity to the antibacterial insect defensins. However, it has marked similarities to the plant defensins and γ-thionins, which are both defence molecules. Six drosomycin-like molecules have been identified in *D. melanogaster* and one in diapausing adults of *D. triauraria*. *D. triauraria* drosomycin has 70% homology to the one from *D. melanogaster*. Drosomycin is a potent antifungal peptide affecting the growth of filamentous fungi including human and plant pathogens at a micromolar level. In contrast to heliomicin, an antifungal peptide from the tobacco budworm *Heliothis virescens*, drosomycin has no effect on yeast strains. Surprisingly, among the large panel of bacterial strains tested, none of them is sensitive to drosomycin. At a high salt concentration, both peptides retain their biological efficacy, unlike insect defensins. Recently, Thevissen and co-workers established that *Rs*AFP2, a plant antifungal peptide from radish seed (*Raphanus sativus*) with sequence similarities to drosomycin and heliomicin, interacts with fungal glucosylceramides in an

Defensin:

Def, Defensin, 2R mapping 46D9

*MKFFVLVAIAFALLACVAQA*QPVSDVDPIPEDHVLVHEDAHQEVLQHSRQKR**ATCDLLSKWNWNHTACAGHCIAKGFKGG YCNDKAVCVCRN**-92

Drosomycins:

Drs, Drosomycin, 3L mapping 63D2-63D2

*MMQIKYLFALFAVLMLVVLGANEADA***DCLSGRYKGPCAVWDNETCRRVCKEEGRSSGHCSPSLKCWCEGC**-70

Dro2, Drosomycin-like D, 3L mapping 63D1-63D1

MHGGGVVLYKRPHKLSIPNVFENSSKVPDSFGTSVAMVQIKFLFVFLAVMTIVVLAANMADA**DCLSGKYKGPCAVWDNEM CRRICKEEGHISGHCSPSLKCWCEGC**

Dro3, Drosomycin-like E (G in SwissProt Database), 3L mapping 63D1-63D1

*MVQMIFLFAILAVMTIVLMEANTVLAR***DCLSGTFGGPCWAWSGEKCRRLCIEEGHVSGHCSGAMKCWCEGC**-71

Dro4, Drosomycin-like F, 3L mapping 63D1-63D1

*MAQIKGLFALLAVVTIVLMVANSASAV***DCPSGRFSGPCWAWDGEQCRRLCREEGRVSGHCSASLKCWCEQC**-71

Dro5, Drosomycin-like B-G (E in SwissProt Database), 3L mapping 63D1-63D1

*MQIKFLYLFLAVMTIFILGAKEADA***DCLSGRYGGPCAVWDNETCRRVCKEEGRSSGHCSPSLKCWCEGC**-69

Dro6, Drosomycin-like I (D in SwissProt Database), 3L mapping 63D1-63D1

*MMQIKFLFTFLALLMMVILGAKEADA***DCLSGRYRGPCAVWDNETCRRVCREEGRGRVSGHCSARLQCWCEGC**-72

Drs-l, Drosomycin-like H, 3L mapping 63D1-63D1

*MVQIKFLFVFLAVMTIVVLAANMADA***DCLSGKYKGPCAVWDNEMCRRICKEEGHISGHCSPSLKCWCEGC**-70

Metchnikowins:

Mtk, Metchnikowin, 2R, 52A1

*MQLNLGAIFLALLGVMATATSVLAEP***HRHQGPIFDTRPSPFNPNQPRPGPIY**-52

------------------*T*-------**HRRQGPIFDTRPSPFNPNQPRPGPIY**

Fig. 2. The *Drosophila* antifungal drosomycins (seven genes *Drs*, *Dos2–6*, *Drs-l*) and metchnikowins (two allelic forms originating from a single *Mtk* gene), and the *Drosophila* defensin (one gene *Def*). The primary structure of the precursors (the size is indicated at the end of each sequence) is reported with the one letter code, the signal peptides are in *italics*, and the mature peptides in **bold**. Gene location and chromosomal location are also mentioned.

initial step of the process of fungal growth inhibition [10]. While this interaction is not by itself sufficient, it is nevertheless necessary for inducing fungal growth arrest. As far as the mode of action of antifungal defensins is concerned, breakthrough studies are yet to surface. Mature drosomycin, present in the hemolymph at an approximate concentration of 100 μ*M*, has been characterized through differential analysis by reversed-phase HPLC from an acidic extract of experimentally infected flies versus control flies (without any screening of in vitro antimicrobial assays) and cDNA cloning experiments. The intronless *drosomycin* gene (*Drs*), which encodes this variant of drosomycin, maps on chromosome 3L at position 63D2, and is coding a 70-residue precursor polypeptide containing a signal peptide (26 amino acids) followed by the mature drosomycin. Six additional *drosomycin* genes (*Dro2–6*, *Drs-l*) are located also on chromosome 3L but are mapping at position 63D1 (fig. 2). Very recently, a differential analysis by microreversed-phase HPLC and MALDI-TOF mass spectrometry of hemolymph collected from naïve flies versus experimentally infected flies evidenced the presence of a N-glycosylated (supposed to be composed of two N-acetylhexosamines, three hexoses and one pentose) drosomycin with a primary sequence corresponding to the one of the *Drs* gene product.

Metchnikowin, A Cysteine-Free *Drosophila* Antifungal Peptide

Metchnikowin is a linear 26-amino acid proline-rich (27%) antifungal peptide without cysteine residues. This peptide has primary structure similarities (38%), including a consensus Pro-Arg-Pro sequence, to the C-terminal domain of the largest proline-rich antibacterial peptides from insects namely to abaecin from the bumblebee *Bombus pascuorum* and to lebocins 1–3 from the silk moth *Bombyx mori*. Metchnikowin has also marked similarities to short proline-rich insect AMPs such as drosocin [11]. Two forms of metchnikowin, which differ by one amino acid in position three (histidine/arginine), have been isolated and characterized from an acidic extract of experimentally infected flies through in vitro screening of antimicrobial activity, and cDNA cloning. During the process of its purification, metchnikowin, present at an estimated circulating concentration of 10–20 μ*M*, was reported to affect the growth of the filamentous fungus *Neurospora crassa* and of the gram-positive bacterium *Micrococcus luteus*. Nevertheless, recombinant metchnikowin was found to inhibit exclusively the growth of filamentous fungi and for this reason metchnikowin is now classified as a strictly antifungal peptide. The intronless *metchnikowin* gene (*Mtk*), which maps on chromosome 2R at position 52A1, is coding a 52-residue precursor polypeptide containing a signal peptide (26 amino acids) followed by the mature metchnikowin (fig. 2). The presence of two isoforms of metchnikowin in the acidic extract performed on several hundreds of experimentally infected flies and a single gene in the *Drosophila* genome suggests an allelic polymorphism in the population used.

Cecropins, α-Helical Peptides Largely Distributed in Higher Insect Orders

Cecropins and cecropin-like are 29–42 residues in size, linear and cysteine-free AMPs. Cecropins were the first animal inducible AMPs to be isolated and fully characterized and are one of the landmarks of the AMPs produced by lepidopteran and dipteran insects in response to an experimental infection [4]. The first insect cecropin has been isolated from the blood of experimentally infected diapausing pupae of the moth *Hyalophora cecropia* (Lepidoptera) and the first genes coding *Drosophila* AMPs were the *cecropin* genes. Primary structure comparisons revealed that dipteran cecropins form a particularly homologous group with an identity ranging from 70% up to 100% as observed between cecropin IA/sarcotoxin IA from *S. peregrina* and cecropin A from *Drosophila* [4]. By homology to the solution conformation of *Hyalophora* cecropin demonstrated by circular dichroism and NMR spectroscopy in water/hexafluoropropanol mixtures, *Drosophila* cecropin may have a long N-terminal, basic, amphipathic α-helix and a shorter and more hydrophobic C-terminal helix, linked by a short hinge region (fig. 1) [4]. Synthetic *Drosophila* cecropin has a high efficacy on a large panel of gram-negative bacteria at concentrations below 10 μ*M*, while most of the gram-positive strains tested remained insensitive even at higher concentrations. Interestingly, some recent reports established that cecropins might also affect the growth of some filamentous fungi and yeasts. Interestingly, structure-activity relationship (SAR) studies on animal AMPs are dominated by data on cecropins and on magainins (α-helical AMPs from frog skin secretions). It is speculated that the helix-forming capability of these molecules, in contact with the lipidic components of the bacterial membrane, results in a general disintegration of the membrane structure and lysis of the bacteria. Cecropins A-C are synthesized as 63-residue precursors with a putative signal-peptide of 23 residues followed by mature cecropin (39 residues) ending by a glycine residue that is participating to the C-terminal amidation of the *Drosophila* cecropins (fig. 3). Only the *Drosophila* cecropin A (at an estimated circulating concentration of 20 μ*M*) has been detected, isolated and structural characterized through a differential display analysis by microreversed-phase HPLC and MALDI-TOF mass spectrometry between blood samples collected from naïve *Drosophila* and experimentally infected flies. Interestingly, four *cecropin* genes (*CecA1-A2*, *CecB* and *CecC*) plus a pseudo gene (*Cec-ψ1*), located on chromosome 3R at position 99E2, were found in the *Drosophila* genome.

Drosocin, An Anti-Gram-Negative O-Glycopeptide

Drosocin is a 19-residue cysteine-free peptide with a high proportion in proline residues (>30%) organized in three repeats Pro-Arg-Pro segments. Drosocin (fig. 1) is the prototype of the short-sized O-glycosylated

Drosocin:

Dro, Drosocin, 2R mapping 51C1

*MKFTIVFLLLACVFAMGVATP***GKPRPYSPRPTSHPRPIRV**RREALAIEDHLTQAAIRPPPILPA-64

Cecropins:

CecA1-A2, Cecropin A, 3R mapping 99E2-99E2

*MNFYNIFVFVALILAITIGQSEA***GWLKKIGKKIERVGQHTRDATIQGLGIAQQAANVAATAR**G-63

CecB Cecropin B, 3R mapping 99E2-99E2

*MNFNKIFVFVALILAISLGNSEA***GWLRKLGKKIERIGQHTRDASIQVLGIAQQAANVAATAR**G-63

CecC,Cecropin C, 3R mapping 99E2-99E2

*MNFYKIFVFVALILAISIGQSEA***GWLKKLGKRIERIGQHTRDATIQGLGIAQQAANVAATAR**G-63

Cec-ψ*1*, Pseudogen, 3R mapping 99E2-99E2

*MKDNTIWILFVLILTTYWPQLEA***GFDGTLDVGLYTLNATLQVAEVASKAANVAITAR**G-58

Attacins and MPAC:

AttA, Attacin A, 2R mapping 51C1-51C1

*MQKTSILIVALVALFAITEALPS*LPTTGPIRVRR**QVLGGSLTSNPAGGADARLDLTKGIGNPNHNVVGQVFAAGNTQSGPVTTGGTLAYNNAGHGASLTKTHTPGVKDVFQQEA HANLFNNGRHNLDAKVFASQNKLANGFEFQRNGAGLDYSHINGHGASLTHSNFPGIGQQLGLDGRANLWSSPNRATTLDLTGSASKWTSGPFANQKPNFGAGLGLSHHF**G-224

AttB, Attacin B, 2R mapping 51C1-51C1

*MQKTSILILALFAIAEA*VPTTGPIRVRR**QVLGGSLASNPAGGADARLNLSKGIGNPNHNVVGQVFAAGNTQSGPVTTGGTLAYNNAGHGASLTKTHTPGVKDVFQQEAHANLFN NGRHNLDAKVFASQNKLANGFEFQRNGAGLDYSHINGHGASLTHSNFPGIGQQLGLDGRANLWSSPNRATTLDLTGSASKWTSGPFANQKPNFGAGLGLSHHF**G-218

AttC, Attacin C and MPAC, 2R mapping 50A3-50A3

*MSKIVLLIVVIVGVLGSLAVA*LP**QRPYTQPLIYYPPPPTPPRIYRA**RR**QVLGGSLTSNPSGGADARLDLSKAVGTPDHHVIGQVFAAGNTQTKPVSTPVTSGATLGYNNHGHGL ELTKTHTPGVRDSFQQTATANLFNNGVHNLDAKAFASQNQLANGFKFDRNGAALDYSHIKGHGATLTHANIPGLGKQLELGGRANLWQSQDRNTRLDLGSTASKWTSGPFKGQT DLGANLGLSHYF**G-241

AttD, Attacin D, 3R mapping 90B6-90B6

MECQASGNPKSGAATAQCGVRVGDDLANARAGVFASTPGAGGPVTKGVYGAVNANGHALSLQHGHIEGVGSTTTAAAQANLFQSNNAALNATAFHSHSRSHDQFGGGLNLQTGT GHQAAVGVTRVPQFGMTAVQASGTANLYTSPSGNLSLNATGSANHHLRGPMRGKSDFGTGVNLRYNF-181

Diptericins:

Dpt, Diptericin, 2R mapping 55F8-55F8

*MQFTIAVALLCCAIASTLAYPMP***DDMTMKPTPPPQYPLNLQGGGGGQSGDGFGFAVQGHQKVWTSDNGRHEIGLNGGYGQHLGGPYGNSEPSWKVGSTYTYRFPNF**-106

Dpt B, DiptericinB, 2R mapping 55F8-55F8

IQLTKPNQIKMHFTASLLFIGLACAFSSAWAYPYPDPREIVNLQPEPLAYAPNFDVPLHRVRRQFQLNGGGGGSPKQGFDLSLNGRAPVWQSPNGRHSFDATGSYAQHLGGPYG NSRPQWGAGGVYTFRF-130

Fig. 3. The *Drosophila* AMPs that are preferentially efficient on gram-negative bacteria: drosocin (O-glycopeptide, one gene *Dro*), cecropins (C-terminally amidated, three genes *CecA1-A2*, *CecB*, *CecC* and a pseudo gene *Cec-*ψ*1*), attacins (four genes *AttA-D*), and diptericins (two genes *Dpt* and *DptB*). MPAC that is acting synergistically with cecropins corresponds to the maturated pro-domain of attacin C (**underlined bold** sequence in the attacin C precursor). The primary structure of the precursors (the size is indicated at the end of each sequence) is reported with the one letter code, the signal peptides are in *italic*, and the mature peptides in **bold**. The underlined glycine residue (C-terminus) is involved in the process of C-terminal amidation of the cecropins and attacins. Gene location and cytogenetic mapping are indicated.

proline-rich AMPs that have been isolated from insects [6, 11]. Strong primary structure similarities exist for example between drosocin and apidaecins from the honeybee *Apis mellifera* (details on similarities between the short-chain proline-rich AMPs from insects are available in [11]). Circular dichroism and two-dimensional NMR spectroscopy studies reveal that there is no substantial difference in the random coil conformations observed between glycosylated drosocin and its non-glycosylated analogue. Nevertheless, some subtle differences in the small populations of folded conformers, suggest a more extended structure within the glycosylation site. Although a series of different glycoforms (averaging a circulating concentration of 40 μM) have been isolated from the hemolymph of experimentally infected *Drosophila* using in vitro antimicrobial activity screening, no sugar-free drosocin was detected in such a test. Further analysis showed that the presence of the sugar moiety on Thr^{11} is necessary for the full biological activity of drosocin. Nevertheless this is not a general conclusion as for the antibacterial O-glycopeptide pyrrhocoricin, isolated from the European sap-sucking bug *Pyrrhocoris apterus*, the absence of the sugar motif improves the efficacy of the molecule on the bacteria [11]. Recently, a differential display analysis by mass spectrometry has identified, in addition to the drosocin substituted on Thr^{11} by one or two sugars, a doubly substituted form of drosocin bearing an additional O-glycan motif on Ser^7 [6]. This suggests that drosocin may be synthesized by the fat body of experimentally infected flies in a more complex form than initially reported. The in vitro activity of drosocin and analogues has been extensively studied. Remarkably, drosocin has a high selectivity toward gram-negative bacteria especially the ones belonging to the *Enterobacteriaceae* family, while gram-positive strains remain mostly non-susceptible. SAR studies on drosocin and other short-chain proline-rich AMPs, highlight some interesting biological properties. Unlike the α-helical AMPs and most of the cysteine-stabilized AMPs that kill bacteria within minutes through non-stereospecific interactions, drosocin takes several hours to kill bacteria. In addition, the all-D-drosocin is totally ineffective suggesting that drosocin is bactericidal through a mechanism that includes stereoselective elements and not a pore-forming or detergent-like process. Using a drosocin/pyrrhocoricin model, Otvos and co-workers observed that shortening these peptides from either termini or cyclizing the molecule in a head-to-tail fashion destroys their activity. Using biotin- and fluorescein-labelled peptides, they reported that pyrrhocoricin and certainly drosocin interact with the 70-kDa *E. coli* heat shock protein DnaK. This suggests that drosocin binds to the multi-helical lid above the substrate-binding pocket of *E. coli* DnaK resulting in the inhibition of chaperone-assisted protein folding [12]. As prominent sequence variations exist in the binding region of eukaryotic and prokaryotic DnaK molecules, short-chain proline-rich AMPs may represent promising starting points to design strain-specific antibiotics lacking any activity on mammalian cells. Interestingly, drosocin is synthesized as a 64-residue precursor corresponding to a signal-peptide (21 residues) followed by the mature drosocin (12 residues) and ending by a pro-domain of 24 residues (fig. 3). This precursor polypeptide is encoded by an intronless gene (*Dro*), which maps on chromosome 2R at position 51C1.

Diptericin, An Anti-Gram-Negative O-Glycopeptide

The *Drosophila* diptericin cDNA was cloned by homology to *Protophormia* diptericins (A and B) that have been isolated through biochemical approaches. *Drosophila* diptericin has marked similarity to the *Protophormia* diptericin (>60%), to *Sarcophaga* diptericin/sarcotoxin (approximately 54%) and to *Glossina marsitans* diptericin (>50%). Only very recently mature *Drosophila* diptericin has been identified in the blood of experimentally infected flies by microreversed-phase HPLC and MALDI-TOF mass spectrometry, and by an immunological assay using an anti-*Drosophila*-diptericin antibody. *Drosophila* diptericin is an 83-mer peptide with a mass measured by mass spectrometry higher than the one calculated from the primary sequence deduced from the cDNA. The presence of an O-glycosylation on a threonine residue within the N-terminal proline-rich domain of *Drosophila* diptericin was evidenced by endoproteinase digestion followed by MALDI-TOF mass spectrometry. Such a structural modification is reminiscent of the O-glycosylation observed on drosocin and MPAC (see below) as well as on *Protophormia* diptericin. In addition to carrying two complex carbohydrate side-chains, *Protophormia* diptericin boasts a C-terminally amidated amino acid, a modification that could not be observed in *Drosophila* diptericin as no extended glycine residue exists at the C-terminus of the *Drosophila* diptericin precursor. By homology to *Protophormia* and *Sarcophaga* diptericins, *Drosophila* diptericin may be active on a panel of gram-negative bacteria below the 10 μM range. To perform some SAR studies, synthetic versions of unglycosylated and O-glycosylated full size 82-mer *Protophormia* diptericin were prepared by solid-phase synthesis [13] and/or expressed in *Escherichia coli*, respectively [14]. The full-size synthetic *Protophormia* diptericin was found to have an activity for only a limited number of gram-negative bacteria and to kill these bacteria within an hour by increasing the permeability of the outer and inner membranes of *Escherichia coli* strain D22. NMR spectroscopy studies and circular dichroism (CD) measurements revealed that diptericin is lacking any defined structure even in a lipid environment such as liposomes. *Drosophila* diptericin is synthesized as a 106-residue precursor corresponding to a signal-peptide (23 residues) directly followed by the mature diptericin (fig. 3). This precursor polypeptide is encoded by an intronless gene (*Dpt*), which maps on chromosome 2R at position 55F8. A second partially sequenced *diptericin* gene (*DptB*) has been identified in the *Drosophila* genome and is also located on chromosome 2R at position 55F8.

Attacins, Large Polypeptides with Antibacterial Properties

Among the *Drosophila* AMPs, attacins are the largest molecules with more than 190 residues. If we except the presence of a C-terminal glycine residue that may participate to a C-terminal amidation, attacins, which are cyteine-free, have no particular post-translational modifications. Attacins were initially reported from Lepidoptera and their presence in *Drosophila* has only been evidenced by cDNA cloning experiments and PCR-based differential display. *Drosophila* attacins have marked similarities (55%) to attacins from *G. morsitans*. Less than 40% similarity exists between the *Drosophila* attacins and their counterparts in *S. peregrina* and *P. terranovae*, and in *H. cecropia* and *B. mori*, two dipteran and lepidopteran insects, respectively. The rather large size of this polypeptide renders activity studies difficult and for this reason only a limited number of information is available. By homology to the *Hyalophora* attacin A, the *Drosophila* attacins A-C may target more effectively gram-negative bacteria. No estimation of the attacin concentration in *Drosophila* blood is reported because of the poor detection of these peptides by a biochemical approach. Nevertheless, considering the stability of attacin C in the blood of immune-challenged flies identical to that of MPAC (see below), the concentration of attacin C may be estimated as similar to that of MPAC. *Hyalophora* attacin was found to interfere with transcription of the *omp* gene in the representative gram-negative strain *E. coli*. The *omp* gene is involved in the synthesis of porines, which form protein channels in membranes. This results in a breakdown of the external membrane of *E. coli*. Four *attacin* genes (*AttA-D*) were found in the *Drosophila* genome. Three genes are located on chromosome 2R with two (*AttA* and *B*) at position 51C1 (adjacent to the *drosocin* gene) and one (*AttC*) at position 50A3. The fourth gene (*AttD*) is located on chromosome 3R at position 90B6, and deduction of the primary structure of mature attacin D remains difficult to be defined precisely.

MPAC, the *Drosophila* Attacin C Pro-Domain with Antibacterial Properties

MPAC (fig. 1) is a cysteine-free 23-residue proline-rich peptide (35% proline residues) post-translationally modified form of the pro-domain of the antibacterial polypeptide attacin C (maturated pro-domain of attacin C, see figure 3 underlined bold sequence) [6]. This peptide has been detected, isolated and structurally characterized through a combination of biochemical approaches, namely a differential display analysis either by microreversed-phase HPLC and MALDI-TOF mass spectrometry of blood samples collected from naïve

Drosophila versus experimentally infected flies. Compared to the pro-domain of attacin C (see fig. 3), MPAC is lacking N- and C-terminal dipeptides, bears a N-terminus pyroglutamic acid, and carries an O-glycosylation motif on Thr^{16}. Such a structural motif is reminiscent of the structure architecture of the proline-rich Drosocin (fig. 1, 2). Amino acid comparisons show that MPAC has marked similarities (37%) with the pro-domains of other diptericins from *D. melanogaster* and *D. simulans* and less than 18% identity with lepidopteran diptericins. Slightly less than 40% identity exists between MPAC and other short-chain proline-rich AMPs from insects such as drosocin and heliocin, a 22-residue peptide that was isolated from the lepidopteran *Heliothis virescens* (direct submission to SwissProt under the accession number P83427). Heliocin also shares with MPAC a N-terminally blocked pyroglutamate and an O-glycosylated threonine [6]. The strongest similarity (46%) was observed between MPAC and the N-terminal proline-rich domain of the flesh fly *S. peregrina* diptericin/sarcotoxin IIA. Due to evident similarities to drosocin and other proline-rich AMPs from insects, functional analysis of MPAC was evaluated in vitro using synthetic non-glycosylated and glycosylated forms against a large variety of bacteria. Surprisingly, none of the strain sensitive to drosocin appeared to be sensitive to MPAC, while a significant and representative synergy was observed between MPAC and cecropin [for details, see ref. 7]. The circulating concentration in MPAC was estimated to be 5 or 1 μM following an infection with the gram-negative strain *Enterobacter cloacae* or the gram-positive strain *M. luteus*, respectively [6]. MPAC is synthesized as a 241-residue precursor corresponding to a signal-peptide (21 residues) followed by MPAC (23 residues) downstream a dibasic cleavage site that will generate mature attacin C (fig. 3) ending by a glycine residue that is participating to the C-terminal amidation of attacin C. This precursor polypeptide is encoded by an intronless gene (*AttC*), which maps on chromosome 2R at position 50A3.

Rel Proteins Control Expression of Antimicrobial Peptide Genes

The genes encoding most AMPs in *Drosophila* are strongly induced or upregulated in response to infection. Classical deletion mapping of the promoters performed in the early 1990's on the promoters of the *cecropin A1* and *diptericin* genes led to the identification of infection response elements. Deletion or mutations of these sequences strongly reduces or abrogates inducibility in tissue culture transfected cells or in transgenic flies in vivo. Strikingly, the sequences independently identified in the *cecropin A1* and *diptericin* promoters (GGGGATTYYT) are closely related and evoke the DNA motif recognized by members of the Rel family of transcription factor, to which NF-κB belongs. Such Rel binding sites were subsequently identified in the promoters of all other AMP genes [1]. The *Drosophila* genome encodes three members of the Rel family. The first one, Dorsal, was identified in a mutagenesis screen for genes involved in the differentiation of the dorsoventral axis in early embryos. In the absence of this transcription factor, the ventral most cells fail to form mesoderm, and embryos die as hollow tubes of dorsal cuticle. Although still expressed at later developmental stages and in adults, and induced by septic injury, Dorsal does not appear to play a critical role in the regulation of the immune response, as *dorsal* mutant larvae or flies exhibit normal responses to infectious challenge. Dorsal may, however, participate in the control of AMP gene expression in larvae (see below).

The *Dif* (for Dorsal-related immunity factor) gene was isolated shortly after the importance of Rel binding sites in the promoters of the AMP genes was recognized. The DIF factor exhibits 48% identity with Dorsal in the Rel domain, and its gene is localized within 10 kb of Dorsal on the left arm of the second chromosome. DIF is not expressed or required in early embryos, unlike Dorsal, however, it is essential for the inducibility of a subset of genes in flies challenged with gram-positive bacteria or fungi. In particular, *drosomycin* induction by these micro-organisms is strongly reduced in *Dif* mutant flies. The induction of *defensin* is also affected. Curiously, induction of these genes is not affected in *Dif* mutant larvae because Dorsal can substitute for DIF. Like Dorsal, the 70 kDa DIF factor is maintained in the cytoplasm of resting cells by the ankyrin repeat IκB-like inhibitor Cactus (fig. 4).

The *Drosophila* genome encodes a third member of the Rel family, named Relish, which is strongly induced in infected flies. Like mammalian p105 and p100, Relish has a composite structure, containing an N-terminal Rel domain, linked to C-terminal IκB-like ankyrin repeats. Upon septic injury, Relish is rapidly processed by endoproteolysis to generate two stable fragments, REL-68, which contains the Rel homology domain and translocates to the nucleus, and REL-49, which contains the ankyrin repeats and remains cytosolic. Cleavage occurs at a caspase target site, and may be mediated by the caspase Dredd, which is required to induce expression of the genes encoding antibacterial peptides, and interacts physically with Relish in transfected tissue-culture cells [15].

The Toll and IMD Pathways Control Inducible Expression of AMP Genes

Two distinct pathways, Toll and IMD, trigger activation of Dif and Dorsal on one hand, and Relish on the other hand. The Toll pathway, named after the

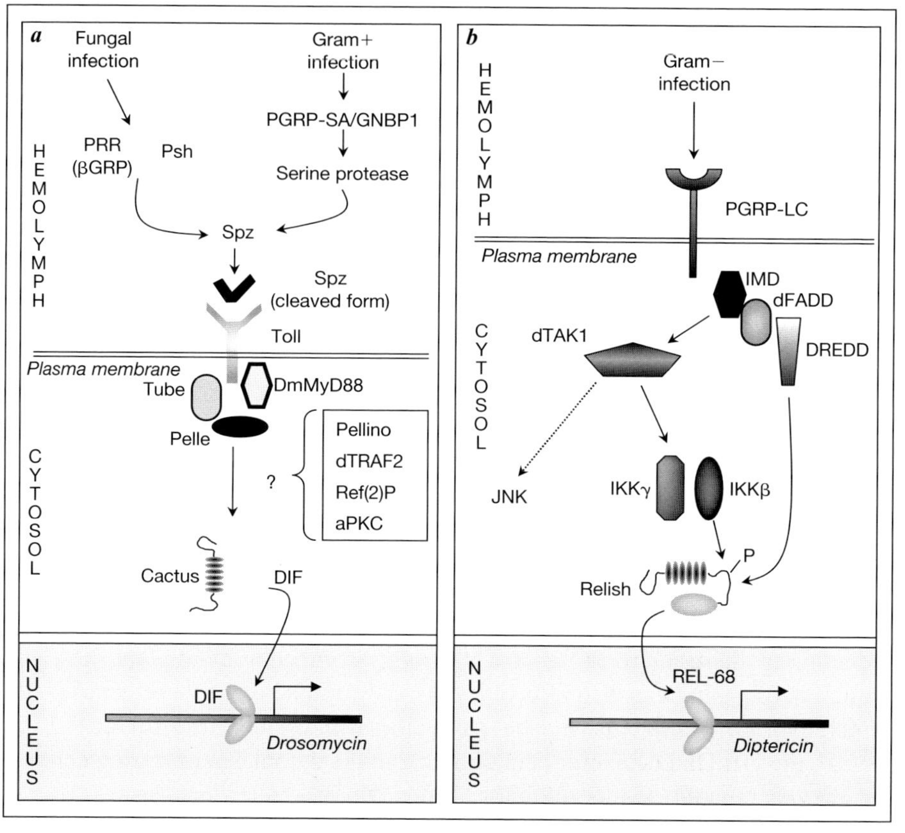

Fig. 4. Induction of genes coding AMPs by the Toll and IMD pathways. ***a*** Infection by fungi and gram-positive bacteria trigger the Toll pathway. Microbes are recognized by PRRs (PGRP-SA and GNBP1 for gram-positive bacteria, presumably a β-glucan recognition protein (βGRP) for fungi). Recognition of infectious non-self activates a proteolytic cascade involving serine-proteases such as the *persephone* gene product (PSH), and results in the cleavage of the cytokine Spaetzle (Spz) to generate an active Toll ligand. Toll activation leads to phosphorylation and degradation of the inhibitor Cactus, and nuclear translocation of the transcription factor DIF. ***b*** Peptidoglycan from gram-negative bacteria activates the membrane receptor PGRP-LC. IMD acts downstream of PGRP-LC and activates Relish by IKKβ-mediated phosphorylation, and endoproteolytic cleavage, most likely by the caspase DREDD. The 68-kDa Rel domain from Relish then translocates to the nucleus where it can induce expression of antibacterial peptide genes. See text for details and references.

transmembrane receptor Toll and initially characterized genetically for its role in establishment of the dorso-ventral polarity of the embryo, controls phosphorylation and subsequent degradation of Cactus thus liberating the Rel proteins Dorsal or Dif. On the other hand, the IMD pathway, named after the first mutant of the antibacterial response, *immune deficiency*, controls processing of Relish [1].

To date, three types of stimuli have been shown to activate the Toll pathway: (1) during development, a proteolytic cascade involving the products of the genes *snake* and *easter* is activated on the ventral side of the embryo. The last serine-protease of the cascade, Easter, cleaves the cysteine-knot growth factor Spaetzle to release a stable 12-kDa C-terminal fragment, which binds to Toll with nanomolar affinity and activates signalling [16]; (2) in adults flies, recognition of infection with gram-positive bacterial infections appears to involve two distinct pattern recognition receptors (PRRs), the peptidoglycan (PGN) recognition protein PGRP-SA encoded by the gene *semmelweiss* (*seml*) [17], and the β-glucan recognition (βGRP)/Gram-negative binding protein (GNBP)1, encoded by the gene *osiris* [18]. These genes are thought to then activate serine proteases distinct from those active in the embryo, which cleave and activate Spz, and (3) recognition of fungal infections involves different PRRs, as *seml* and *osi* mutant flies exhibit wild-type responses to this type of infection. These PRRs in turn activate serine proteases, one of which is encoded by the gene *persephone*, and also trigger processing of the Spaetzle protein to generate an active Toll ligand [19].

Upon activation, the Toll receptor triggers phosphorylation of Cactus. This inhibitory protein is subsequently polyubiquitinated and degraded by the 26S proteasome. The kinase phosphorylating Cactus has not been identified yet, but is different from the IκB kinase (IKK) complex. At the plasma membrane level, Toll interacts with a complex of three death domain containing proteins, DmMyD88, Tube and the serine/threonine kinase Pelle. The factors Pellino, dTRAF2, atypical PKC and the scaffolding protein Ref(2)P, which is structurally related to mammalian p62, have been suggested to participate in signal transduction between Pelle and Cactus. However, these findings have not yet been fully tested in vivo using mutant flies. In response to infection by gram-positive bacteria or fungi, the Toll pathway triggers expression of the antifungal peptides drosomycin and metchnikowin, as well as dozens of other genes, many of which remain to be characterized.

Infection by gram-negative bacteria triggers the distinct IMD pathway. Diaminopymelic (DAP)-type PGN from the cell wall of these bacteria activates the transmembrane type II receptor PGRP-LC [20]. Three different splicing isoforms of PGRP-LC, LCa, LCx and LCy, are expressed, which differ in the PGRP domain present in the extracellular domain. Recent data indicate that these isoforms specifically recognize different forms of PGN: whereas polymeric DAP-type PGN signal through PGRP-LCx, monomeric forms of this PGN, such as tracheal cytotoxin (TCT; a monomeric disaccharide-tetrapeptide fragment of PGN), require both PGRP-LCx and PGRP-LCa to trigger a response in *Drosophila* cells. The stimulus activating PGRP-LCy has not been identified yet. The situation at the receptor level is probably more complex, as null mutants of the *PGRP-LC* gene exhibit less severe phenotypes than null mutants of other genes in the pathway, such as *relish*, *kenny*, or *ird5*, suggesting that another receptor can to some extent substitute for PGRP-LC. In addition, overexpression of a distinct member of the PGRP family, PGRP-LE, which has also been shown to interact with PGN from gram-negative bacteria, has been shown to trigger expression of *diptericin* in the absence of infection. The secreted molecule PGRP-LE may therefore act as a coreceptor of PGRP-LC for gram-negative PGN.

Upon activation, the PGRP-LC receptor activates the death domain signalling molecule encoded by the *imd* gene, which interacts through its death domain with the adaptor dFADD. dFADD in turn interacts with and activates the caspase DREDD. dFADD also activates the kinase dTAK1, which is thought to phosphorylate and activate the IKK complex, composed of the catalytic subunit IKKβ encoded by the gene *kenny* and the regulatory subunit IKKγ encoded by *ird5*. *Drosophila* IKKβ phosphorylates Relish, which is preassociated with DREDD, presumably triggering processing of the transcription factor [1].

Other Signaling Pathways Activated during the *Drosophila* Immune Response

A large number of genes in addition to those encoding AMPs are induced or upregulated by infection. Microarray analysis revealed that, in addition to the Toll and IMD pathways, the Jun kinase (JNK) and JAK/STAT pathways are also activated during the immune response, and contribute to the upregulation of several genes in response to microbial challenges [21].

In particular, immune challenge leads to the rapid and transient upregulation of many genes encoding components of the cytoskeleton (e.g. *myosin31DF*, *actin88F*, *flightin*), regulators of the cytoskeleton (e.g. *ninjurin*) and proapoptotic factors (e.g. *reaper*). The JNK pathway controls expression of these molecules. Interestingly, the JNK pathway is connected to the same upstream signalling cassette than the Relish-pathway. In particular, activation of JNK requires Imd and TAK1. Thus, there is a branching point in the signalling pathway triggered by gram-negative bacteria downstream of TAK1, with one branch regulating cytoskeletal regulators through the JNK pathway, and another branch regulating AMP expression through the NF-κB related molecule Relish [21] (fig. 4). The rapid and transient induction of JNK in response to septic injury may be connected to tissue repair processes.

Another group of proteins are expressed transiently but later in the course of infection. These genes, which encode secreted molecules such as turandot, or thiol-ester containing proteins (TEPs), are induced at a maximal level 6 h after infection [21]. They are regulated by the JAK/STAT pathway, which controls expression of acute phase proteins in response to cytokines such as interleukin-6 in mammals. Characterization of the role of the JAK/STAT signalling in the *Drosophila* immune response has revealed a previously unsuspected role for blood cells in the orchestration of some aspects of the systemic response to infection in flies.

Hemocytes in the *Drosophila* Immune Response

Blood cells have long been known to participate in some aspects of host-defence reactions in flies. In particular, natural infections of larvae by parasitoid wasp eggs triggers differentiation of a specialized blood cell type, the lamellocyte, which encapsulates the eggs. However, this response does not involve the induction of AMPs. More recently, the importance of blood cells has been recognized in the control of bacterial infection. For example, *domino* mutant larvae, in which multiplication of diploid cells is affected, have a severely reduced number of hemocytes, and exhibit increased sensitivity to bacterial infection. Similarly, inhibition of phagocytosis in *Drosophila* blood cells by injection of polystyrene beads into the body cavity sensitises adult flies to infection by *E. coli*. In addition to phagocytosis, hemocytes also have important signalling functions. Indeed, natural infection per os of *domino* mutant larvae with the gram-negative bacterium *Erwinia carotovora* fails to induce *diptericin* expression in the fat body, suggesting that hemocytes detect infection and signal to the fat body. Whereas induction of AMPs in the fat body after septic injury is not dependent on hemocytes, induction of other target genes is critically dependent on hemocytes. This is for example the case of turandot (Tot) A. Induction of TotA in fat body cells requires both the IMD pathway, and the JAK/STAT pathway. Upon septic injury with gram-negative bacteria, hemocytes are stimulated to produce the cytokine Unpaired-3 (Upd-3), which activates the receptor domeless (dome) on fat body cells. Domeless, which is structurally related to the gp130 subunit of the interleukin-6 receptor, in turn activates the JAK kinase encoded by the gene *hopscotch*, and the STAT factor encoded by *marelle* [22]. Altogether, these data reveal that in the course of infection, the fat body can integrate signals from different origins, ranging from microbial molecules (IMD pathway) to hemocyte-derived cytokines (JAK/STAT pathway). Apart from fat body cells and blood cells, epithelial cells also play an active role in antimicrobial defences.

Epithelial Responses in *Drosophila*

Important physiological functions such as nutrient absorption, reproduction, gas exchange, or excretion necessitate interaction between host cells and the environment. As a result, epithelial cells from the digestive, reproductive, respiratory tracts, or from the excretory system are frequently exposed to micro-organisms. Local expression of AMPs in these tissues plays an important role as a first line of defence in mammals but also in insects, and in particular in *Drosophila*. Interestingly, epithelial expression of AMPs follows a complex pattern that is specific for each peptide [23] (fig. 5a). For example, *defensin* and *metchnikowin* are expressed in two small glands that open at the beginning of the alimentary canal, the labellar glands; *drosomycin* is expressed in the salivary glands, *diptericin*, *attacin* and to a lesser extent *drosocin* and *metchnikowin* are expressed in the midgut, *diptericin*, *cecropin* and *metchnikowin* are expressed in the Malpighian tubules (the excretory system) and *cecropin*, *defensin*, *drosomycin*, *drosocin*, and to a lesser extent *metchnikowin* and *attacin* are expressed in specific parts of the male and female reproductive tracts. In larvae, *cecropin* expression can also be

a

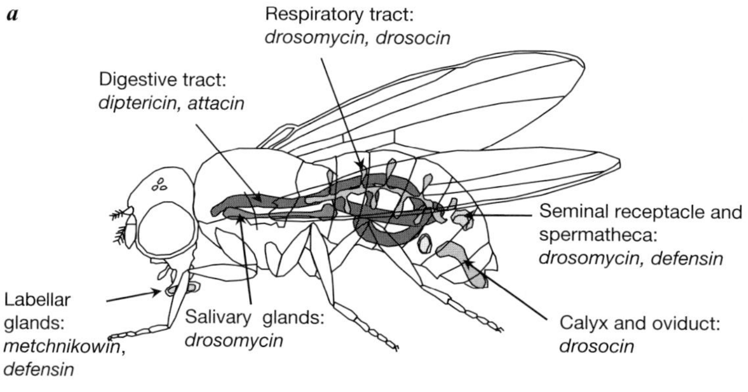

b

Tissue	Expression	Regulation
Fat body	Inducible	Toll
Trachea	Inducible	IMD
Salivary gland	Constitutive	Caudal
Spermatheca	Constitutive	?

Fig. 5. Epithelial expression of antimicrobial peptides in *Drosophila*. ***a*** The main sites of expression of AMP genes in a female fly are shown. ***b*** Tissue-specific expression and regulation of the *drosomycin* gene.

induced in the epidermis [24]. It would be interesting in the future to investigate epithelial expression of isoforms of AMPs, for example the *cecropins* or the *drosomycins*, to see if the members of the family are expressed in the same tissues, or in different subsets of tissues. In most cases, epithelial expression of AMPs is not constitutive, and is only observed in a restricted area of the tissue, in a fraction of the flies, suggesting a response to a local infection. Natural infection of the flies or larvae with the Gram-negative bacteria *E. carotovora* triggers inducible expression of AMPs in several surface epithelia in a tissue-specific manner. In all cases, this induction is dependent on the IMD pathway [23, 24]. In particular, *drosomycin* expression, which is regulated by the Toll pathway during the systemic response, is regulated by *imd* in the respiratory tract, thus demonstrating the existence of distinct regulatory mechanisms for local and systemic induction of AMPs in *Drosophila*. In some tissues, expression of the AMPs is constitutive. This includes the salivary glands for *drosomycin*, the female reproductive tract for *drosomycin*, *drosocin*, *defensin* and *cecropin*, and the male reproductive tract for *cecropin*. The constitutive expression of *drosomycin* in salivary glands and *cecropin* in the ejaculatory duct was recently shown to be dependent on the homeobox gene product Caudal [25]. Curiously, the constitutive expression of *drosomycin* in the female reproductive tract is not dependent on Caudal (fig. 5b).

Another interesting feature of the epithelial expression of AMPs is that it reveals an additional level of complexity in their genetic regulation. Indeed, the fact that AMP genes like *diptericin* and *drosocin*, which are both controlled exclusively by the IMD pathway in the fat body during the systemic response, are induced in different epithelial locations (e.g. digestive tract versus tracheae), points to the existence of tissue-specific transcription factors which probably act in concert with Relish to mediate induction of a subset of peptides in a given epithelium [23]. The existence of important co-factors for Relish is supported by the fact that the GATA transcription factor encoded by the gene *serpent* modulates the inducible expression of some AMPs (e.g. cecropin) in larval fat body [26]. The *cecropin A1* (but also *cecA2*, *cecB* and *defensin*) promoter contains a conserved motif named R1 in the close vicinity of the Rel binding site that contributes to the high level of AMP gene expression and may contribute to the tissue-specific expression of these genes [27]. The identity of the transcription factor binding to this motif is not known at this stage.

Concluding Remarks

In conclusion, AMPs are potent effectors of the *Drosophila* immune response. *Toll* or *imd* mutant flies, in which induction of AMP synthesis is impaired exhibit high susceptibility to fungal and/or bacterial infections.

Furthermore, transgene-mediated constitutive expression of AMPs, alone or in combination, restores (at least to some extent) resistance to infection in *Toll* or *imd* mutant flies [28]. In addition to their critical antimicrobial role, the *Drosophila* AMPs have provided geneticists with valuable markers to monitor the immune response and screen for genes controlling recognition of infectious non-self and signal transduction. In spite of the important progress made in the past decade on the characterization of *Drosophila* AMPs, several important questions remain, such as (1) the precise mode of action of these natural antibiotics; (2) the function and tissue distribution of the AMP isoforms revealed by the sequencing of the *Drosophila* genome, and (3) the regulation of the tissue-specific expression of AMP in barrier epithelia.

Acknowledgements

We thank Dr R. Stöcklin for critical reading and useful comments on the manuscript, and Dr Jules A. Hoffmann for continued interest and support.

References

1 Hoffmann J: The immune response of Drosophila. Nature 2003;426:33–38.
2 Tossi A, Sandri L: Molecular diversity in gene-encoded, cationic antimicrobial polypeptides. Curr Pharm Des 2002;8:743–761.
3 Pag U, Sahl HG: Lanthionine-containing bacterial peptides; in Dutton CJ, Haxell MA, McArthur HAI, Wax RG (eds): Peptide Antibiotics: Discovery, Modes of Action, and Applications. New York, Marcel Dekker, 2002, pp 47–80.
4 Bulet P, Stöcklin R, Menin L: Anti-microbial peptides: From invertebrates to vertebrates. Immunol Rev 2004;198:169–184.
5 Ganz T, Lehrer B: Antimicrobial peptides; in Ezekowitz RAB, Hoffmann JA (eds): Innate Immunity. Infectious Diseases. Totowa, Humana Press, 2003, pp 287–303.
6 Rabel D, Charlet M, Ehret-Sabatier L, Cavicchioli L, Cudic M, Otvos L Jr, Bulet P: Primary structure and in vitro antibacterial properties of the Drosophila melanogaster attacin C Pro-domain. J Biol Chem 2004;279:14853–14859.
7 Bulet P, Charlet M, Hetru C: Antimicrobial pepties in insect immunity; in Ezekowitz RAB, Hoffmann JA (eds): Innate Immunity. Infectious Diseases. Totowa, Humana Press, 2003, pp 89–107.
8 Cociancich S, Ghazi A, Hetru C, Hoffmann JA, Letellier L: Insect defensin, an inducible antibacterial peptide, forms voltage-dependent channels in *Micrococcus luteus*. J Biol Chem 1993; 268:19239–19245.
9 Garcia-Olmedo F, Molina A, Alamillo JM, Rodriguez-Palenzuela P: Plant defense peptides. Biopolymers 1998;47:479–491.
10 Thevissen K, Warnecke DC, Francois IE, Leipelt M, Heinz E, Ott C, Zahringer U, Thomma BP, Ferket KK, Cammue BP: Defensins from insects and plants interact with fungal glucosylceramides. J Biol Chem 2004;279:3900–3905.
11 Bulet P, Hetru C, Dimarcq JL, Hoffmann D: Antimicrobial peptides in insects; structure and function. Dev Comp Immunol 1999;23:329–344.
12 Kragol G, Lovas S, Varadi G, Condie BA, Hoffmann R, Otvos L Jr: The antibacterial peptide pyrrhocoricin inhibits the ATPase actions of DnaK and prevents chaperone-assisted protein folding. Biochemistry 2001;40:3016–3026.

13 Cudic M, Bulet P, Hoffmann R, Craik DJ, Otvos L Jr: Chemical synthesis, antibacterial activity and conformation of diptericin, an 82-mer peptide originally isolated from insects. Eur J Biochem 1999;266:549–558.

14 Winans KA, King DS, Rao VR, Bertozzi CR: A chemically synthesized version of the insect antibacterial glycopeptide, diptericin, disrupts bacterial membrane integrity. Biochemistry 1999;38:11700–11710.

15 Stoven S, Silverman N, Junell A, Hedengren-Olcott M, Erturk D, Engstrom Y, Maniatis T, Hultmark D: Caspase-mediated processing of the Drosophila NF-kappaB factor Relish. Proc Natl Acad Sci USA 2003;100:5991–5996.

16 Weber AN, Tauszig-Delamasure S, Hoffmann JA, Lelievre E, Gascan H, Ray KP, Morse MA, Imler JL, Gay NJ: Binding of the Drosophila cytokine Spatzle to Toll is direct and establishes signaling. Nat Immunol 2003;4:794–800.

17 Michel T, Reichhart JM, Hoffmann JA, Royet J: Drosophila Toll is activated by Gram-positive bacteria through a circulating peptidoglycan recognition protein. Nature 2001;414:756–759.

18 Gobert V, Gottar M, Matskevich AA, Rutschmann S, Royet J, Belvin M, Hoffmann JA, Ferrandon D: Dual activation of the Drosophila toll pathway by two pattern recognition receptors. Science 2003;302:2126–2130.

19 Ligoxygakis P, Pelte N, Hoffmann JA, Reichhart JM: Activation of Drosophila Toll during fungal infection by a blood serine protease. Science 2002;297:114–116.

20 Kaneko T, Goldman WE, Mellroth P, Steiner H, Fukase K, Kusumoto S, Harley W, Fox A, Golenbock D, Silverman N: Monomeric and polymeric gram-negative peptidoglycan but not purified LPS stimulate the Drosophila IMD pathway. Immunity 2004;20:637–649.

21 Boutros M, Agaisse H, Perrimon N: Sequential activation of signaling pathways during innate immune responses in Drosophila. Dev Cell 2002;3:711–722.

22 Agaisse H, Petersen UM, Boutros M, Mathey-Prevot B, Perrimon N: Signaling role of hemocytes in Drosophila JAK/STAT-dependent response to septic injury. Dev Cell 2003;5:441–450.

23 Tzou P, Ohresser S, Ferrandon D, Capovilla M, Reichhart JM, Lemaitre B, Hoffmann JA, Imler JL: Tissue-specific inducible expression of antimicrobial peptide genes in Drosophila surface epithelia. Immunity 2000;13:737–748.

24 Onfelt Tingvall T, Roos E, Engstrom Y: The imd gene is required for local Cecropin expression in Drosophila barrier epithelia. EMBO Rep 2001;2:239–243.

25 Ryu JH, Nam KB, Oh CT, Nam HJ, Kim SH, Yoon JH, Seong JK, Yoo MA, Jang IH, Brey PT, Lee WJ: The homeobox gene Caudal regulates constitutive local expression of antimicrobial peptide genes in Drosophila epithelia. Mol Cell Biol 2004;24:172–185.

26 Tingvall TO, Roos E, Engstrom Y: The GATA factor Serpent is required for the onset of the humoral immune response in Drosophila embryos. Proc Natl Acad Sci USA 2001;98:3884–3888.

27 Uvell H, Engstrom Y: Functional characterization of a novel promoter element required for an innate immune response in Drosophila. Mol Cell Biol 2003;23:8272–8281.

28 Tzou P, Reichhart JM, Lemaitre B: Constitutive expression of a single antimicrobial peptide can restore wild-type resistance to infection in immunodeficient Drosophila mutants. Proc Natl Acad Sci USA 2002;99:2152–2157.

Jean-Luc Imler, PhD
CNRS UPR9022, Institut de Biologie Moléculaire et Cellulaire
15, rue René-Descartes, FR–67084 Strasbourg Cedex (France)
Tel. +33 388 417 036, Fax +33 388 606 926, E-Mail JL.Imler@ibmc.u-strasbg.fr

Kabelitz D, Schröder J-M (eds): Mechanisms of Epithelial Defense.
Chem Immunol Allergy. Basel, Karger, 2005, vol 86, pp 22–41

Antimicrobial Peptides in Human Skin

Jürgen Harder, Jens-Michael Schröder

Clinical Research Unit at the Department of Dermatology, University Hospital Schleswig-Holstein, Campus Kiel, Kiel, Germany

Abstract

Human skin is permanently exposed to a wide variety of potential harmful microorganisms. Despite these microbial threats, skin is surprisingly highly resistant against infections. Various studies in the last decade discovered a chemical cutaneous defense system based on the production of antimicrobial proteins. These antimicrobial proteins act as a first defense line through their broad spectrum of potent antimicrobial activity. In addition to constitutively expressed antimicrobial proteins, production of various antimicrobial proteins in keratinocytes is induced by bacterial compounds as well as proinflammatory cytokines. The resulting local accumulation of antimicrobial proteins offers a fast and very efficient way to prevent microbes from establishing an infection. In this review we summarize the current status of our knowledge of expression of antimicrobial proteins in human skin.

Introduction

Skin represents the major epithelium of man and provides a barrier to the external environment. Given that the skin is permanently threatened by various potential pathogenic micro-organisms, it is very plausible that the epidermis would express chemical substances to prevent microorganism penetration, in addition to the physical barrier of the lipid-rich stratum corneum.

In 1987, Michael Zasloff showed that frog skin protects itself by the release of antimicrobial peptides called magainins [1]. These small endogenous gene-encoded antimicrobial proteins exhibit a broad-spectrum of antimicrobial activity. These observations gave rise to the assumption that human skin may also produce antimicrobial proteins which contribute to innate cutaneous defense.

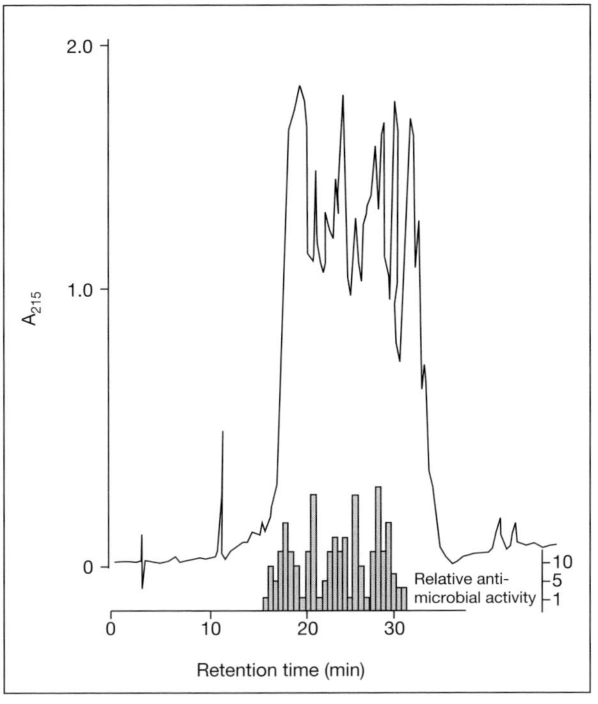

Fig. 1. C8-reversed-phase HPLC of heparin-affinity column bound stratum corneum proteins. Healthy donor-derived heparin-bound stratum corneum proteins were separated by preparative C8 reversed-phase HPLC and proteins were recorded at 215 nm absorbance. 30 μl aliquots of HPLC-fractions were tested for *E. coli* bactericidal activity using the radial diffusion assay system and the diameter of clearing zones was expressed (bars).

Analysis of human psoriatic skin extracts as well as heel stratum corneum extracts revealed that human skin harbors a broad set of antimicrobial proteins active against various bacteria and fungi. Figure 1 shows an example of a reversed-phase C8 high pressure liquid chromatography (HPLC) separation of heparin-bound proteins that have been isolated from a stratum corneum extract. Analysis of the resulting fractions for antimicrobial activity against the gram-negative bacterium *E. coli* revealed high antimicrobial activities in various fractions (fig. 1). This indicates that human skin has the ability to produce a broad spectrum of antimicrobial proteins.

In recent years the discovery of several skin-derived antimicrobial proteins has brought new insights into the role of antimicrobial proteins for cutaneous host defense. In the following, we summarize the current status of our knowledge of expression of antimicrobial peptides/proteins in human skin.

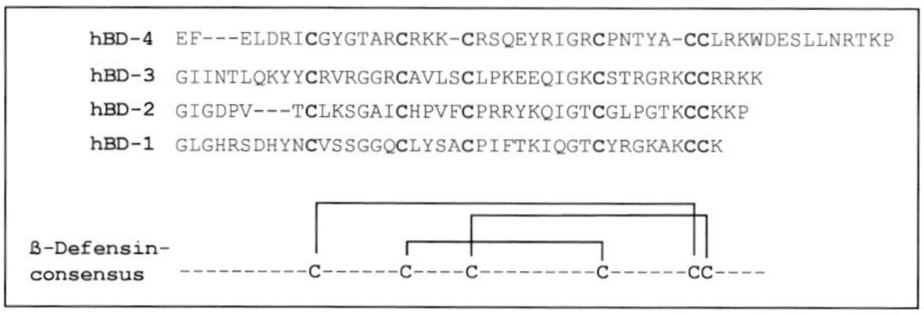

Fig. 2. Amino acid sequences and alignment of human β-defensins including the connectivity of the putative disulfide bonds.

β-Defensins

Human β-Defensin-2

HBD-2 is a small, cationic peptide of 43 amino acids belonging to the class of β-defensins. These small antimicrobial peptides share a unique disulfide-bonding motif of six conserved cysteines (fig. 2). HBD-2 was originally isolated from lesional psoriatic scale extracts using an *Escherichia coli* affinity column [2]. In contrast to psoriatic skin, healthy skin extracts harbor only low amounts of hBD-2 peptide [3]. The high expression of hBD-2 in psoriasis was confirmed by in situ hybridization and immunohistochemistry demonstrating strong staining for hBD-2 mRNA and peptide in lesional psoriatic keratinocytes [4, 5]. As yet it is not fully understood why lesional psoriatic scales contain such high amounts of hBD-2. However, it seems likely that hBD-2 is induced by the actions of increased levels of proinflammatory cytokines present in psoriatic skin, because hBD-2 was found to be the first inducible member of the human defensin family. For induction of hBD-2 in skin keratinocytes either endogenous proinflammatory cytokines like IL-1α, IL-1β, TNF-α or exogenous stimuli like bacteria such as *Pseudomonas aeruginosa* have proven to be the most effective hBD-2-inducers [2, 4, 6, 7]. Interestingly, a mucoid clinical isolate of *P. aeruginosa* that originated from an ulcus was found to be a very potent and effective inducer of hBD-2 expression in respiratory epithelial cells [8]. Similar results we obtained in cultured primary keratinocytes [own unpubl. results]. The bacterial factors (pathogen-associated molecular patterns, PAMPs) as well as the appropriate receptors on keratinocytes (pattern recognition receptors, PRRs) involved in the bacteria-mediated hBD-2 induction have not yet been identified. Recently, it has been shown that human keratinocytes express various Toll-like receptors (TLRs) and that TLR-2 mediates NF-κB dependent gene expression in keratinocytes stimulated with *Staphylococcus aureus* and its cell wall components [9]. These data could indicate that TLRs may play a role for the bacteria-mediated induction of antimicrobial proteins in keratinocytes. Whereas LPS was found to be a poor inducer of hBD-2 expression in keratinocytes, it has been suggested that keratinocyte response to bacterial LPS is rather indirect in vivo: Mononuclear phagocytes respond towards LPS by release of IL-1β, which was identified as the most powerful endogenous inducer of hBD-2 production in keratinocytes [10].

Interestingly, a recent study indicated, that keratinocytes can discriminate between commensal and pathogenic bacteria: Induction of hBD-2 in foreskin-derived keratinocytes by the skin commensal *Staphylococcus epidermidis* was suppressed by inhibitors of the JNK and p38 pathways, whereas induction by the skin pathogen *Streptococcus pyogenes* was blocked by inhibitors of nuclear factor kappa B (NF-κB) [11].

The inducible gene expression of hBD-2 was found to be in concordance with the features of its promoter region: The hBD-2 promoter contains several putative transcription factor-binding sites, including nuclear factor-κB (NF-κB), activator protein 1 (AP-1), AP-2 and NF-IL-6, which are known to be involved in the induction and regulation of inflammatory responses [5, 8].

HBD-2 expression is variable in skin, where it is more readily detectable in facial skin and foreskin than skin from abdomen and breast. HBD-2 transcripts are found in keratinocytes within interfollicular skin. Interindividual and site-specific differences in intensity of immunostaining were observed and the pattern of peptide localization was seen to be rather focal [12]. Ultrastructural analyses using immunogold labelling detected hBD-2 in lamellar bodies and intercellular spaces of IL-1α-stimulated cultured primary keratinocytes. These findings indicate that hBD-2 is released with the lipoid compounds stored preformed in lamellar bodies and that hBD-2 can accumulate in intercellular spaces of the uppermost skin layers where it could reach due to the very small inter space volume a high local concentration that effectively could control microbial growth [13].

In normal skin, hBD-2 immunoreactivity is localized to the uppermost layers of the epidermis and/or stratum corneum [12], which is in concordance with recent studies showing an upregulation of hBD-2 in cultured primary keratinocytes brought to differentiation by high calcium concentrations [14, 15]. The increased expression of hBD-2 in more differentiated keratinocytes confirms the hypothesis that the highest amounts of hBD-2 are produced in the more differentiated terminal skin layers, thus providing a first line defense system to invading microbes. In concordance with these data, it has been reported that in an organotypic epidermal culture increasing concentrations of hBD-2 peptide were detected in the more differentiated suprabasal layers and that calculated concentrations of hBD-2 in IL-1α-induced epidermal cultures reached concentrations within the antimicrobial range of hBD-2 (3,5–16 μ*M*). Because of the restricted pattern of hBD-2 distribution in the epidermis, its local concentration is likely to be higher [7].

Natural hBD-2 has been reported to exhibit primarily antimicrobial activity against gram-negative bacteria such as *E. coli* and both nonmucoid and mucoid strains of *P. aeruginosa* (concentration necessary to kill 90% bacteria (LD_{90}) = 10 μg/ml). Natural hBD-2 exhibits less activity against yeast such as *Candida albicans* (LD_{90}, 25 μg/ml) and only bacteriostatic activity against *S. aureus* at concentrations greater than 100 μg/ml [2, 3]. Similar to natural hBD-2, in insect cells generated recombinant hBD-2 exhibits broad spectrum antimicrobial activity against skin-relevant microorganisms like *P. aeruginosa, S. epidermidis* and *C. albicans*, but only bacteriostatic activity against *S. aureus* [7]. Activity of hBD-2 against gram-negative bacteria depends on ion composition and is sensitive to increasing concentrations of NaCl, where its activity is diminished [8, 16–18]. The capacity of hBD-2 to kill bacteria in vivo has been demonstrated in a mouse gene therapy study with hBD-2-transfected tumor cells. In this study, hBD-2-bearing tumors were associated with fewer viable bacteria than controls following bacterial infection [19].

In addition to its capacity to serve as an peptide antibiotic, hBD-2 provides a link between innate and adaptive immunity. HBD-2 shows selective chemotactic activity for cells stably transfected with human CCR6, a G-protein-coupled chemokine receptor preferentially expressed by immature dendritic cells and memory T cells. Thus, hBD-2 may promote adaptive immune responses by recruiting dendritic and T cells to the site of microbial invasion through interaction with CCR6 [20]. This hypothesis has been further supported by structural analysis of human macrophage inflammatory protein-3α/CCL20, the natural ligand for CCR6, and comparison with the structure of hBD-2, where marked similarities are seen [21]. Very recently, novel chemotactic properties of hBD-2 have been described. HBD-2, but not hBD-1, was found to be a specific chemoattractant for TNF-α-treated human neutrophils [22]. The effect of hBD-2 on neutrophils was G-protein-phospholipase C pathway-dependent and ligand-receptor analysis indicated that the activity is mediated by CCR6. HBD-2 is also a chemotaxin for mast cells via a pertussis toxin-sensitive and G protein-phospholipase C-dependent pathway that does not utilize CCR6 [23]. Furthermore, hBD-2 stimulates histamine release from these cells and induces prostaglandin D synthesis, suggesting a role in the recruitment of mast cells to inflammation foci and in the modulation of proinflammatory mast cell functions [24].

Human β-Defensin-3

Based on the observation that hBD-2 is not able to kill the gram-positive germ *S. aureus*, it has been hypothesized that human skin produces apart from hBD-2 other antimicrobial peptides acting against *S. aureus*. In consequence of this hypothesis, in one study a 5-kDa antimicrobial peptide has been isolated from lesional psoriatic scales using an biochemical approach. This peptide was identified as a novel third member of the human β-defensin family and accordingly termed hBD-3. The complete cDNA of hBD-3 was cloned from keratinocytes and lung epithelial cells [25]. In two parallel studies, hBD-3 was cloned based on bioinformatics and functional genomic analysis [26, 27]. The cDNA sequence of hBD-3 encodes for a 67 amino acid peptide precursor that contains a 22 amino acid signal peptide, is approximately 43% identical to hBD-2 and shares the β-defensin six cysteine motif [25–27] (fig. 2).

The isolation and purification from lesional psoriatic scales, molecular cloning and biochemical analysis of antimicrobial peptides in cell culture supernatants has revealed keratinocytes and airway epithelial cells as cellular sources of hBD-3. HBD-3 is expressed throughout epithelia of many organs and in some non-epithelial tissues. Skin, gingival keratinocytes, tonsils, esophagus, trachea, placenta, adult heart, skeletal muscle and fetal thymus comprise the major hBD-3 mRNA-expressing tissues [25–27].

In cultured primary keratinocytes, TNF-α induces hBD-3 mRNA expression at a low level [25, 28]. Interestingly, however, interferon-γ, which doesn't induce hBD-2, is a more powerful hBD-3-inducing cytokine in keratinocytes [25, 26, 28]. Like bovine epithelial β-defensins and hBD-2, hBD-3 is also induced by contact of keratinocytes or tracheal epithelial cells with bacteria [25, 26].

Synthetic and recombinant hBD-3, which are indistinguishable from the naturally occurring peptide with respect to their antimicrobial activity and biochemical properties, exhibit a broad spectrum of potent antimicrobial activity against many potentially pathogenic gram-negative and gram-positive bacteria and fungi, including multiresistant *S. aureus* (MRSA) and vancomycin-resistant *Enterococcus faecium* (VRE) [25, 26, 29–31]. Testing 44 *S. aureus* clinical isolates for susceptibility to hBD-3, including 22 MRSA strains, revealed that 55% of the MRSA strains exhibited more than 20% survival in the presence of 1 μg/ml hBD-3. Only three (13%) of the methicillin-sensitive *S. aureus* isolates exhibited more than 20% survival with hBD3. Furthermore, inactivation of the fmtC gene in *S. aureus*, which is associated with methicillin resistance, results in increased susceptibility of MRSA to hBD-3 [32].

The mechanism of hBD-3 microbicidal activity is not yet known. Ultrastructural investigation of hBD-3-treated *S. aureus* has revealed signs of perforation of the peripheral cell wall, with explosion-like liberation of the plasma membrane within 30 min and bacteriolysis after 2 h. The morphological effects resemble those seen when *S. aureus* is treated with penicillin [25].

Human hBD-3 does not fold preferentially into a native conformation in vitro under various oxidative conditions. Chemically synthesized topological analogs of hBD-3 with predefined disulfide connectivities, including the (presumably) native β pairing, have revealed similar antimicrobial activity for differently folded hBD-3 species, indicating that hBD3 antimicrobial activity is independent of disulfide connectivity. Interestingly, hBD-3 dependent chemotaxis of monocytes and CCR6-transfected HEK 293 cells strongly depend on the topology of disulfide connectivities in hBD3. The absence of any disulfide bridge abolishes the chemotactic activity of hBD-3, suggesting that a defined 3D structure stabilized by disulfide bonding is required for productive binding to and activation of CCR6-receptor. In contrast to chemotatic activity, the bactericidal activity of hBD-3 remains unaffected by the absence of any disulfide bridge [33].

Human β-Defensin-1

hBD-1 is the first discovered human β-defensin. It was originally isolated from human blood filtrate as a novel peptide with significant sequence homology to bovine β-defensins [34]. Mature hBD-1 is a small, cationic peptide of 36 amino acid residues containing six cysteines forming three intramolecular disulfide bonds (fig. 2). In blood plasma and urine, several forms of hBD-1 have been isolated ranging in length from 36 to 47 amino acid residues and differing from each other by amino terminal truncation [35]. Natural hBD-1 protein has not yet been isolated from human skin. However, using in situ hybridization Fulton et al. [36] detected hBD-1 mRNA expression in the suprabasal keratinocytes and sweat ducts of human skin. A more detailed investigation of hBD-1 expression in human skin by in situ hybridization and immunohistochemistry confirmed the expression of hBD-1 in keratinocytes. HBD-1 mRNA and peptide were found to be consistently expressed in skin samples from various body sites and localized to suprabasal keratinocytes within interfollicular skin. HBD-1 transcripts and peptides were also detected in sweat glands and sebaceous glands [12]. In vitro induction of keratinocyte differentiation by calcium treatment led to the upregulation of hBD-1 gene expression [15, 37, 38]. This may explain why hBD-1 peptide shows a stronger expression in more differentiated terminal layers of human skin (Malpighian layer, stratum corneum) [12].

Overexpression of hBD-1 in keratinocytes resulted in increased expression of differentiation markers, suggesting that hBD-1 promotes differentiation of keratinocytes [38]. In contrast to hBD-2 and hBD-3, gene expression of hBD-1 in keratinocytes is not markedly induced by proinflammatory cytokines like IL-1β, TNF-α, IFN-γ or by bacteria like *P. aeruginosa* [15].

Only few studies investigated the antimicrobial spectrum of hBD-1. Recombinant and natural hBD-1 forms (36, 39, and 42 amino acids) exhibit salt-sensitive antimicrobial activity against various laboratory and clinical strains of *E. coli* at micromolar concentrations (0.3–10 μ*M*). The 36 amino acid form was found to be antimicrobially active even in normal urine [35]. In addition to activity against *E. coli*, Singh et al. [18] reported that concentrations of 1–10 μg/ml of a recombinant baculovirus-derived hBD-1-preparation kill *P. aeruginosa*. The dose required to kill 50% of *P. aeruginosa* was found to be 1 μg/ml and 100 ng/ml for hBD-1 and hBD-2, respectively. In other studies, only a minor antimicrobial activity of native hBD-1 was detected, resulting in killing of only a few microorganisms [39]. As yet, no studies have reported activity of hBD-1 against pathogenic gram-positive bacteria such as *S. aureus*.

An intragenic polymorphism of the DEFB1 gene that changes a highly conserved Cys to Ser in the peptide coding region has recently been described. The deduced peptide hBD-1(Ser35) cannot form three disulfide bonds, as one of the cysteines is unpaired. Despite a different arrangement of the disulfides, hBD-1(Ser35) revealed similar antimicrobial activity as the normal hBD-1. The authors speculated that the antimicrobial activity likely depends on the ability of hBD-1(Ser35) to adopt an amphipathic conformation in a hydrophobic environment, as suggested by CD spectroscopy [40].

As also shown for hBD-2, hBD-1 exhibits chemotactic activity for cells stably transfected with the chemokine receptor CCR6. As CCR6 is preferentially expressed by immature dendritic cells and memory T cells, these data suggest that hBD-1 like hBD-2 may function, through interaction with CCR6, to recruit immature dendritic cells and memory T cells to cutaneous sites of microbial invasion [20].

Human β-Defensin-4

The fourth member of the human β-defensin family hBD-4 was initially identified by screening the human genome database and [41] (fig. 2). Synthetic hBD-4 revealed antimicrobial activity at micromolar concentrations against *P. aeruginosa* and *Staphylococcus carnosus* and hBD-4 gene expression was upregulated by bacteria in respiratory epithelial cells [41]. Very recently gene expression of hBD-4 was detected in primary keratinocytes [15]. HBD-4 gene expression was moderately upregulated by contact of cultured keratinocytes with bacteria and was strongly induced by PMA (more than 1,000-fold as analyzed by real-time PCR). These data suggest a role of hBD-4 in the innate cutaneous defense system against bacteria. However, so far nothing is known about the expression of hBD-4 peptide in human skin and as yet attempts have failed to isolate hBD-4 peptide from psoriatic scale extracts as well as from healthy human skin-derived stratum corneum extracts. Therefore, further investigations need to be performed to elucidate the expression levels of hBD-4 peptide in skin and the role of hBD-4 in the chemical skin defense system.

Cathelicidin LL-37

LL-37 belongs to the cathelicidin family, a group of antimicrobial and LPS-binding proteins that share a highly conserved 12-kDa N-terminus, termed cathelin, according to a protein isolated from porcine neutrophils [42]. LL-37 is the only member of the cathelin gene family present in the human genome [43]. The gene for the human cathelicidin LL-37 was first identified in a human bone marrow cDNA [44]. It encodes a preproprotein of 18 kDa (human cationic antimicrobial protein with a molecular size of 18 kDa, hCAP-18) which is proteolytically processed by the serine protease proteinase 3 yielding the C-terminal 37-amino acids containing antimicrobial LL-37 peptide [45]. LL-37 exhibits broad spectrum antimicrobial activity in the micromolar range against various gram-negative and gram-positive bacteria as well as fungi [46]. LL-37 is also cytotoxic in the micromolar range to eukaryotic cells under physiological salt conditions and exhibits cytotoxic and antibacterial activities, which are inhibited by human serum, in particular by apolipoprotein A-I [47, 48].

A recent report showed that the N-terminal cathelin-like prosequence of hCAP-18 exhibits antiprotease as well as antimicrobial activity [49]. The authors conclude that after proteolytic cleavage the cathelin-like domain can contribute to innate host defense through inhibition of bacterial growth and limitation of cysteine-proteinase-mediated tissue damage.

LL-37 stimulates chemotaxis for neutrophils, monocytes, and T cells, acting via the formyl peptide-like receptor-1 [50]. Furthermore, Niyonsaba et al. [51] demonstrated that LL-37 induces mast cell chemotaxis possibly through a G_i protein-phospholipase C signalling pathway. These reports indicate that besides its antibacterial activities, LL-37 may have the potential to recruit effector cells to inflammation and infection foci.

LL-37 seems to have the capacity to control inflammation through its ability to reduce the bacteria-induced production of inflammatory mediators by macrophages and epithelial cells [52] thus confirming the role of LL-37 as a multifunctional modulator of innate immune responses.

The expression of LL-37 in skin seems to be tightly regulated because the LL-37 gene is expressed in keratinocytes of inflamed skin but not of healthy skin [53]. Another study demonstrated that LL-37 expression was upregulated in cutaneous injury due to both, synthesis within epidermal keratinocytes and deposition from infiltrated granulocytes [54].

By immunohistochemistry and in situ hybridization LL-37 was localized to both the eccrine gland and sweat ductal epithelial cells. This observation and the antimicrobial activity of LL-37 against various bacteria in the sweat ionic environment indicate that LL-37 may also contribute to the antibacterial activity of human sweat [55]. Very recently, Murakami et al. [56] demonstrated that in sweat LL-37 is processed by a serine protease-dependent mechanism into multiple novel smaller antimicrobial peptides of 20, 30 and 31 amino acids, respectively. These peptides showed increased bactericidal (*E. coli*, *S. aureus*) as well as fungicidal (*C. albicans*) activity and were synergistic, killing bacteria at lower concentrations when present together. Despite enhanced antimicrobial activity, the hemolytic activity of these short antimicrobial peptides decreased relative to that of LL-37.

Interestingly, LL-37 and other antimicrobial proteins like lysozyme and the α-defensins (HNP1–3) were found in the skin and the vernix caseosa of the newborn infant. This observation may indicate a role of endogenous antimicrobial proteins also as innate immune effectors in the skin barrier of the newborn infant [57, 58].

The discovery of various proteins exhibiting antimicrobial activity in vitro raised the question of whether these proteins are able to control bacterial growth in vivo. The relevance of cathelicidins in cutaneous host defence has been demonstrated in a mouse model. Mice deficient in the expression of the cathelicidin CRAMP (the mouse homolog to the human LL-37) were more susceptible to skin infections caused by group A *Streptococcus* (GAS) and GAS mutants resistant to CRAMP produced more severe skin infections in normal mice [59].

Serine Protease Inhibitors Antileukoprotease and Elafin

The human serine protease inhibitor antileukoprotease (ALP; also known as secretory leukoprotease inhibitor, SLPI; mucous protease inhibitor, MPI) is known to be an antiproteolytic compound of human body fluids and mucous secretions [60]. Expression of ALP in human skin has been demonstrated by the isolation of ALP from healthy human stratum corneum and by detection in supernatants of cultured human primary keratinocytes [61]. By Northern-blot analysis constitutive gene expression of ALP could be detected in primary keratinocytes [61]. Whereas treatment of primary keratinocytes with TNF-α and INF-γ for 8 h did not significantly induce ALP gene expression [61], treatment with TNF-α, IL-1β, IGF-1 and TGF-α for 48 h strongly induced ALP mRNA as well as protein expression [6]. This correlates with the observation, that ALP expression is strongly increased in proliferating keratinocytes as typically seen in psoriatic skin and wound healing [62].

In addition to its antiprotease activity, ALP exhibits high antimicrobial activity against a broad range of microorganisms like gram-negative and gram-positive bacteria as well as the yeast *C. albicans*. This suggests that ALP not only protects skin against proteolysis but also controls cutaneous microbial growth [61, 62].

Another skin-derived serine protease inhibitor termed elafin (also known as skin-derived anti-leukoproteinase, SKALP), that is also inducibly expressed in keratinocytes [6, 63–65], has been reported to exhibit antimicrobial activity against *P. aeruginosa* and *S. aureus* [66]. In this study the full-length, 95 amino acids containing synthetic elafin exhibited killing activity against *P. aeruginosa* (93% killing by 2.5 μ*M* elafin) and only weak activity against *S. aureus* (48% killing by 25 μ*M* elafin). The 50 amino acids containing N-terminal domain and the 45 amino acids containing C-terminal domain exhibited lower antimicrobial activity identifying the full-length form as the most active form. Another report demonstrated only growth-inhibiting, but no bactericidal properties of the 57 amino acids containing C-terminal recombinant elafin against three different *P. aeruginosa* strains [64]. This 57 amino-acids containing, truncated elafin form was originally purified from human horny layers [67]. Interestingly, despite its bacteriostatic properties against *P. aeruginosa*, this form exhibited no bacteriostatic activity against *E. coli* [64], a bacterium that is killed in a very effective way by most of the human antimicrobial proteins known so far. These reports indicate that the full-length elafin may be more important as an antimicrobial factor than its truncated forms. Due to its moderate antimicrobial activity found in vitro, it is not clear, how effective elafin functions as an antibacterial agent in the skin. However, the potential of elafin to act as an infection protective factor has been shown in a mouse model, where adenoviral augmentation of elafin protected lungs against injury and infection mediated by *P. aeruginosa* [68].

Dermcidin

Most of the skin-derived antimicrobial proteins are produced by keratinocytes. Recently, it has been shown that also sweat glands have the capacity to produce antimicrobial proteins. Dermcidin is a novel anionic antimicrobial peptide produced and secreted exclusively by human eccrine sweat glands [69]. Dermcidin is proteolytically processed to DCD-1, a 47 amino acids containing peptide exhibiting antimicrobial activity against bacteria (i.e. *E. coli*, *E. faecalis*, *S. aureus*) and *C. albicans*. DCD-1 is found in sweat in antimicrobial active concentrations of 1–10 μg/ml and its antimicrobial activity is not affected by the low pH value and high salt concentrations of human sweat. This indicates that sweat glands may have a function in the innate immune responses of the skin by secreting antimicrobial proteins. As mentioned above, more recent investigations also revealed the presence of the cathelicidin LL-37 and its processed forms in human sweat.

Adrenomedullin

Adrenomedullin is a pluripotent 52-amino-acid peptide having numerous physiological roles, which include vasodilation, renal homeostasis, hormone regulation, neurotransmission, and growth modulation [for review, see 70]. It is expressed by many epithelia including normal and neoplastic skin [71]. Adrenomedullin mRNA and protein expression was detected in keratinocytes of the epidermis and hair follicles, as well as in cells of the eccrine and apocrine sweat glands and sebaceous glands [71].

Adrenomedullin exhibits high antimicrobial activity against *E. coli* (Minimum bactericidal concentration, MBC: 1.5 μg/ml) and moderate activity against *S. aureus* (MBC: 25 μg/ml) [72]. Interestingly, adrenomedullin seems to be very effective in killing *Propionibacterium acnes* (MBC: 25 μg/ml) suggesting that adrenomedullin could play a role in the skin disease acne vulgaris, where *P. acnes* hypercolonization is a common problem. Adrenomedullin is secreted by keratinocytes in vitro (35 fmol/10^6 cells/12 h) but it is not clear whether concentrations within the antimicrobial range of 0.1–10 μg/ml are reached in vivo [72].

In oral epithelial cells as well as in gastric epithelial cells adrenomedullin expression is induced through bacterial challenge [73, 74]. These data provide further evidence that adrenomedullin may play a role in the defense mechanisms initiated by epithelia cells to avoid microbial infection.

Neutrophil Gelatinase-Associated Lipocalin

The neutrophil gelatinase-associated lipocalin (NGAL; also called human neutrophil lipocalin) is a 25-kDa protein initially isolated from the specific granules of human neutrophils [75, 76]. It has been shown that NGAL exhibits bacteriostatic activity through its ability to bind bacterial ferric siderophores [77]. It is known from the antimicrobial protein lactoferrin that its ability to sequester free iron in human body fluids decreases the availability of the metal to invading bacteria thus inhibiting bacterial growth [78]. While lactoferrin simply binds free iron, NGAL specifically binds to ferric siderophores thus inhibiting the siderophore-mediated iron uptake by bacteria. Since iron is essential for bacterial growth, its deprivation causes bacteriostatic effects. Indeed, it has been shown that addition of only 5 μ*M* NGAL to *E. coli* led to a 20-fold growth inhibition [77]. The low expression of NGAL in healthy skin is strongly increased in skin disorders characterized by dysregulated epithelial differentiation such as psoriasis, pityriasis rubra and squamous cell carcinoma [79]. Expression of NGAL in human keratinocytes is upregulated by IL-1β, IGF-I and TGF-α [6].

RNase 7

Keratinocytes also produce RNase 7, a member of the RNase A superfamily, containing ribonuclease as well as antimicrobial activity [80]. RNase 7 can be isolated from heel stratum corneum extracts in quite high amounts (200–500 μg per 50 gram stratum corneum). The contact of keratinocytes with bacteria like *P. aeruginosa* or *S. aureus* increased RNase 7 expression. RNase 7 exhibits a broad spectrum of efficient antimicrobial activity at low micromolar concentrations against various pathogenic microorganisms, including *S. aureus*, *P. aeruginosa*, *P. acnes* and *C. albicans* (lethal doses that achieves a CFU reduction of 90% (LD_{90}) = 0.75–1.5 μ*M*). RNase 7 is also highly active in killing vancomycin-resistant *E. faecium* (LD_{90}, <0.2 μ*M*). Blocking the ribonuclease activity of RNase 7 using a soluble RNase-inhibitor did not reduce the antibacterial activity against *E. coli* [our unpubl. results]. This makes it likely that the ribonuclease activity of RNase 7 is not necessary for its activity against *E. coli*, a characteristic which also has been described for ECP [81].

The capacity of RNase 7 to fragment RNA identifies this molecule as a part of the ribonuclease activities found on human skin, making it necessary to take special precautions when performing experiments with RNA (i.e. by wearing gloves).

Skin Disease Implications

In addition to psoriatic skin, increased levels of antimicrobial proteins were often found in inflamed and/or infected skin areas. For example, the majority of biopsies taken from patients with acne vulgaris displayed a marked upregulation of defensin-2 immunoreactivity in the lesional and perilesional epithelium – in particular in pustules – and a less marked upregulation of defensin-1 immunoreactivity. These data indicate that upregulated β-defensins may be involved in the pathogenesis of Acne vulgaris [82]. Upregulated expression of hBD-2 was also associated with superficial folliculitis [83]. The expression of LL-37 in keratinocytes of patients with condyloma acuminatum and verruca vulgaris was found to be increased suggesting a role of LL-37 in cutaneous infection caused by papillomavirus [84].

Recent investigations associated decreased levels of antimicrobial proteins with burn and chronic wounds. Whereas hBD-2 can be produced by normal skin, it is absent in the full thickness burn wound and burn blister fluid. This suggests a host defense defect within the burn wound and a possible therapeutic role for antimicrobial peptides in the management of burn wounds [85].

Using blocking-antibodies to LL-37, Heilborn et al. [86] demonstrated that LL-37 promotes the re-epithelialization of human skin wounds. The authors speculate that the reduction of LL-37 in chronic wounds impairs re-epithelialization and may contribute to their failure to heal.

As described above, it is well documented that skin of patients suffering from the common chronic inflammatory disease psoriasis, contains elevated amounts of antimicrobial proteins. The increased expression of antimicrobial proteins in psoriasis might be the reason, why patients suffering from psoriasis have less skin infections than expected [87]. In contrast to psoriasis, hBD-2, hBD-3 and LL-37 expression is decreased in acute and chronic lesions from patients suffering from the common chronic inflammatory disease atopic dermatitis [28, 88]. It is believed that the expected induced expression of antimicrobial peptides in the eczematous atopic skin is inhibited as a result of elevated amounts of Th2 cytokines present in atopic skin. Indeed it has been shown, that Th2 cytokines, like IL-4 and IL-13, suppress the cytokine-mediated induction of antimicrobial peptides like hBD-2 and hBD-3 [28, 88]. These data suggest that the low expression of antimicrobial peptides may – at least in part – account for the increased susceptibility of patients with atopic dermatitis to skin infection. Atopic dermatitis is the first skin disease, where a diminished production of antimicrobial proteins correlates with an increased occurrence of skin infections thus highlighting the importance of antimicrobial proteins for skin defence. These observations provoke the speculation that also other recurrent skin infections may be associated with a dysregulated production of antimicrobial proteins.

Conclusion

The widespread occurrence of antimicrobial proteins in various epithelial tissues of diverse organisms indicates that these proteins may play an important role in epithelial innate immunity. Production of antimicrobial proteins provides a rapidly acting defense system that is already engaged before adaptive immune interactions take place. Recent progress in the isolation and characterization of skin-derived antimicrobial proteins revealed, that skin is capable of producing a huge diversity of antimicrobial proteins. Figure 3 summarizes our current knowledge about the expression of antimicrobial proteins in healthy and inflamed skin.

The observation, that many of these antimicrobial proteins are upregulated at sites of infections confirmed the hypothesis, that these antimicrobial proteins might be essential factors to combat against invading microbes. However, a main goal for future research should be to ascertain, whether antimicrobial proteins,

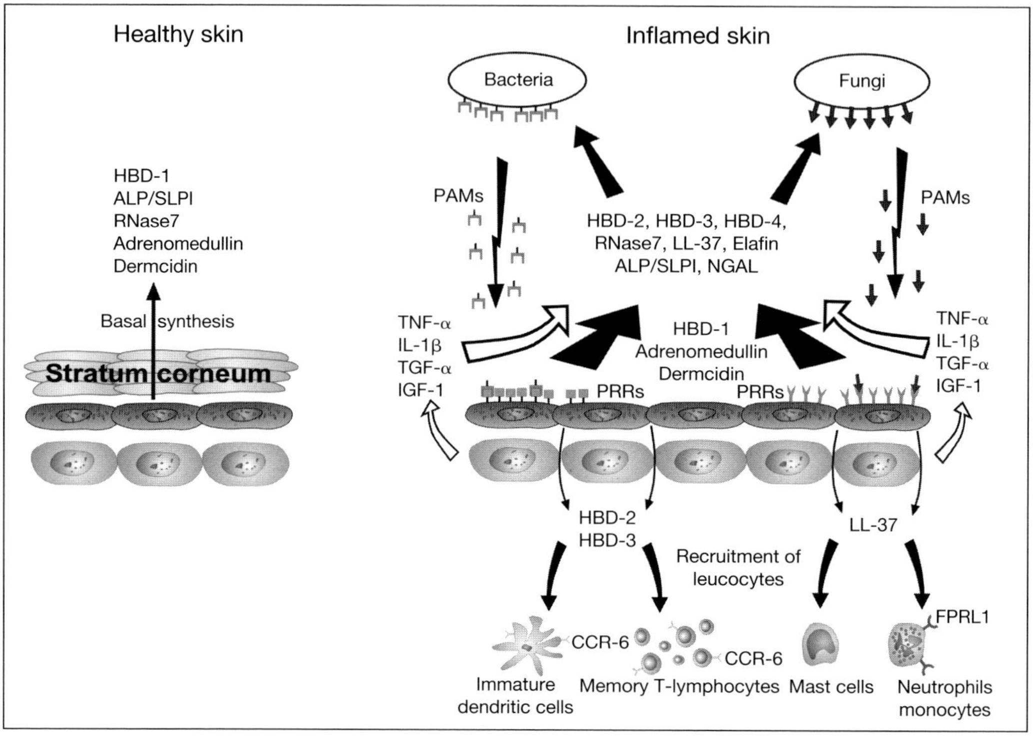

Fig. 3. Expression of antimicrobial proteins in healthy and inflamed skin. Healthy skin is protected through the constitutive expression of antimicrobial proteins derived from keratinocytes and glands (left). Upon disruption of the stratum corneum, which functions as a physical barrier, microorganisms get into contact with the living keratinocytes (right). The keratinocytes recognize microbial surface or secreted compounds ('pathogen associated molecules', PAMs) via special, as yet unknown receptors ('pattern recognition receptors', PRRs) leading to a fast production of inducible antimicrobial proteins, which allows rapid killing of the invading microbes. In addition, the endogenous production of proinflammatory cytokines further increases the expression of antimicrobial proteins at sites of cutaneous infection. Through their chemotactic activities antimicrobial proteins may recruit various leucocytes to the area of infection, thus linking innate immunity with adaptive immunity. β-Defensins may promote adaptive immune responses by recruiting leucocytes through interaction with CCR6 ('CC chemokine receptor 6') whereas LL-37 recruits leucocytes through interaction with FPRL1 ('formyl peptide receptor-like-1') and other as yet unidentified receptors.

which are active in vitro, indeed participate in the prevention of infections in vivo. The problem here is, that as yet we do not have data concerning the local concentrations of antimicrobial proteins in vivo (i.e. for the intercellular space between the keratinocytes where the antimicrobial proteins encounter the microbes). Therefore, it is of substantial interest to investigate, whether the concentrations of antimicrobial proteins reached in vivo are sufficient to exhibit microbicidal properties, keeping in mind that the action of antimicrobial proteins largely depends on the environmental conditions (i.e. ion and protein composition, pH, bacterial physiology). Furthermore, there is currently no information about possible synergistic actions of antimicrobial peptide and -protein mixtures, which most likely will occur in vivo. However, given that the aqueous intercellular space between the keratinocytes in the epidermis is very low, it is likely, that the local concentrations of antimicrobial proteins are high enough to provide a fast killing of invading microorganisms.

References

1 Zasloff M: Magainins, a class of antimicrobial peptides from Xenopus skin: Isolation, characterization of two active forms, and partial cDNA sequence of a precursor. Proc Natl Acad Sci USA 1987;84:5449–5453.

2 Harder J, Bartels J, Christophers E, Schröder J-M: A peptide antibiotic from human skin. Nature 1997;387:861.

3 Schröder J-M, Harder J: Human beta-defensin-2. Int J Biochem Cell Biol 1999;31:645–651.

4 Huh WK, Oono T, Shirafuji Y, Akiyama H, Arata J, Sakaguchi M, Huh NH, Iwatsuki K: Dynamic alteration of human beta-defensin 2 localization from cytoplasm to intercellular space in psoriatic skin. J Mol Med 2002;80:678–684.

5 Liu L, Wang L, Jia HP, Zhao C, Heng HH, Schutte BC, McCray PB Jr, Ganz T: Structure and mapping of the human beta-defensin HBD-2 gene and its expression at sites of inflammation. Gene 1998;222:237–244.

6 Sorensen OE, Cowland JB, Theilgaard-Monch K, Liu L, Ganz T, Borregaard N: Wound healing and expression of antimicrobial peptides/polypeptides in human keratinocytes, a consequence of common growth factors. J Immunol 2003;170:5583–5589.

7 Liu AY, Destoumieux D, Wong AV, Park CH, Valore EV, Liu L, Ganz T: Human beta-defensin-2 production in keratinocytes is regulated by interleukin-1, bacteria, and the state of differentiation. J Invest Dermatol 2002;118:275–281.

8 Harder J, Meyer-Hoffert U, Teran LM, Schwichtenberg L, Bartels J, Maune S, Schröder J-M: Mucoid *Pseudomonas aeruginosa*, TNF-alpha, and IL-1beta, but not IL-6, induce human beta-defensin-2 in respiratory epithelia. Am J Respir Cell Mol Biol 2000;22:714–721.

9 Mempel M, Voelcker V, Kollisch G, Plank C, Rad R, Gerhard M, Schnopp C, Fraunberger P, Walli AK, Ring J, Abeck D, Ollert M: Toll-like receptor expression in human keratinocytes: Nuclear factor kappaB controlled gene activation by *Staphylococcus aureus* is toll-like receptor 2 but not toll-like receptor 4 or platelet activating factor receptor dependent. J Invest Dermatol 2003;121:1389–1396.

10 Liu L, Roberts AA, Ganz T: By IL-1 signaling, monocyte-derived cells dramatically enhance the epidermal antimicrobial response to lipopolysaccharide. J Immunol 2003;170:575–580.

11 Chung WO, Dale BA: Innate immune response of oral and foreskin keratinocytes: Utilization of different signaling pathways by various bacterial species. Infect Immun 2004;72:352–358.

12 Ali RS, Falconer A, Ikram M, Bissett CE, Cerio R, Quinn AG: Expression of the peptide antibiotics human beta defensin-1 and human beta defensin-2 in normal human skin. J Invest Dermatol 2001;117:106–111.

13 Oren A, Ganz T, Liu L, Meerloo T: In human epidermis, beta-defensin 2 is packaged in lamellar bodies. Exp Mol Pathol 2003;74:180–182.

14 Pernet I, Reymermier C, Guezennec A, Branka JE, Guesnet J, Perrier E, Dezutter-Dambuyant C, Schmitt D, Viac J: Calcium triggers beta-defensin (hBD-2 and hBD-3) and chemokine

macrophage inflammatory protein-3 alpha (MIP-3alpha/CCL20) expression in monolayers of activated human keratinocytes. Exp Dermatol 2003;12:755–760.
15 Harder J, Meyer-Hoffert U, Wehkamp K, Schwichtenberg L, Schröder J-M: Differential gene induction of human β-defensins (hBD-1, -2, -3 and -4) in keratinocytes is inhibited by retinoic acid. J Invest Dermatol 2004;in press
16 Bals R, Wang X, Wu Z, Freeman T, Bafna V, Zasloff M, Wilson JM: Human beta-defensin 2 is a salt-sensitive peptide antibiotic expressed in human lung. J Clin Invest 1998;102:874–880.
17 Tomita T, Hitomi S, Nagase T, Matsui H, Matsuse T, Kimura S, Ouchi Y: Effect of ions on antibacterial activity of human beta defensin 2. Microbiol Immunol 2000;44:749–754.
18 Singh PK, Jia HP, Wiles K, Hesselberth J, Liu L, Conway BA, Greenberg EP, Valore EV, Welsh MJ, Ganz T, Tack BF, McCray PB Jr: Production of beta-defensins by human airway epithelia. Proc Natl Acad Sci USA 1998;95:14961–14966.
19 Huang GT, Zhang HB, Kim D, Liu L, Ganz T: A model for antimicrobial gene therapy: Demonstration of human beta-defensin 2 antimicrobial activities in vivo. Hum Gene Ther 2002; 13: 2017–2025.
20 Yang D, Chertov O, Bykovskaia SN, Chen Q, Buffo MJ, Shogan J, Anderson M, Schröder J-M, Wang JM, Howard OM, Oppenheim JJ: Beta-defensins: Linking innate and adaptive immunity through dendritic and T cell CCR6. Science 1999;286:525–528.
21 Hoover DM, Boulegue C, Yang D, Oppenheim JJ, Tucker K, Lu W, Lubkowski J: The structure of human macrophage inflammatory protein-3alpha/CCL20. Linking antimicrobial and CC chemokine receptor-6-binding activities with human beta-defensins. J Biol Chem 2002;277: 37647–37654.
22 Niyonsaba F, Ogawa H, Nagaoka I: Human beta-defensin-2 functions as a chemotactic agent for tumour necrosis factor-alpha-treated human neutrophils. Immunology 2004;111:273–281.
23 Niyonsaba F, Iwabuchi K, Matsuda H, Ogawa H, Nagaoka I: Epithelial cell-derived human beta-defensin-2 acts as a chemotaxin for mast cells through a pertussis toxin-sensitive and phospholipase C-dependent pathway. Int Immunol 2002;14:421–426.
24 Niyonsaba F, Someya A, Hirata M, Ogawa H, Nagaoka I: Evaluation of the effects of peptide antibiotics human beta-defensins-1/-2 and LL-37 on histamine release and prostaglandin D(2) production from mast cells. Eur J Immunol 2001;31:1066–1075.
25 Harder J, Bartels J, Christophers E, Schröder J-M: Isolation and characterization of human beta-defensin-3, a novel human inducible peptide antibiotic. J Biol Chem 2001;276:5707–5713.
26 Garcia JR, Jaumann F, Schulz S, Krause A, Rodriguez-Jimenez J, Forssmann U, Adermann K, Kluver E, Vogelmeier C, Becker D, Hedrich R, Forssmann WG, Bals R: Identification of a novel, multifunctional beta-defensin (human beta-defensin 3) with specific antimicrobial activity. Its interaction with plasma membranes of *Xenopus* oocytes and the induction of macrophage chemoattraction. Cell Tissue Res 2001;306:257–264.
27 Jia HP, Schutte BC, Schudy A, Linzmeier R, Guthmiller JM, Johnson GK, Tack BF, Mitros JP, Rosenthal A, Ganz T, McCray PB Jr: Discovery of new human beta-defensins using a genomics-based approach. Gene 2001;263:211–218.
28 Nomura I, Goleva E, Howell MD, Hamid QA, Ong PY, Hall CF, Darst MA, Gao B, Boguniewicz M, Travers JB, Leung DY: Cytokine milieu of atopic dermatitis, as compared to psoriasis, skin prevents induction of innate immune response genes. J Immunol 2003;171:3262–3269.
29 Hoover DM, Wu Z, Tucker K, Lu W, Lubkowski J: Antimicrobial characterization of human beta-defensin 3 derivatives. Antimicrob Agents Chemother 2003;47:2804–2809.
30 Maisetta G, Batoni G, Esin S, Luperini F, Pardini M, Bottai D, Florio W, Giuca MR, Gabriele M, Campa M: Activity of human beta-defensin 3 alone or combined with other antimicrobial agents against oral bacteria. Antimicrob Agents Chemother 2003;47:3349–3351.
31 Sahly H, Schubert S, Harder J, Rautenberg P, Ullmann U, Schröder J, Podschun R: Burkholderia is highly resistant to human beta-defensin 3. Antimicrob Agents Chemother 2003;47: 1739–1741.
32 Midorikawa K, Ouhara K, Komatsuzawa H, Kawai T, Yamada S, Fujiwara T, Yamazaki K, Sayama K, Taubman MA, Kurihara H, Hashimoto K, Sugai M: Staphylococcus aureus susceptibility to innate antimicrobial peptides, beta-defensins and CAP18, expressed by human keratinocytes. Infect Immun 2003;71:3730–3739.

33 Wu Z, Hoover DM, Yang D, Boulegue C, Santamaria F, Oppenheim JJ, Lubkowski J, Lu W: Engineering disulfide bridges to dissect antimicrobial and chemotactic activities of human beta-defensin 3. Proc Natl Acad Sci USA 2003;100:8880–8885.
34 Bensch KW, Raida M, Magert HJ, Schulz-Knappe P, Forssmann WG: hBD-1: A novel beta-defensin from human plasma. FEBS Lett 1995;368:331–335.
35 Valore EV, Park CH, Quayle AJ, Wiles KR, McCray PB Jr, Ganz T: Human beta-defensin-1: An antimicrobial peptide of urogenital tissues. J Clin Invest 1998;101:1633–42.
36 Fulton C, Anderson GM, Zasloff M, Bull R, Quinn AG: Expression of natural peptide antibiotics in human skin. Lancet 1997;350:1750–1751.
37 Abiko Y, Nishimura M, Kusano K, Yamazaki M, Arakawa T, Takuma T, Kaku T: Upregulated expression of human beta defensin-1 and -3 mRNA during differentiation of keratinocyte immortalized cell lines, HaCaT and PHK16–0b. J Dermatol Sci 2003;31:225–228.
38 Frye M, Bargon J, Gropp R: Expression of human beta-defensin-1 promotes differentiation of keratinocytes. J Mol Med 2001;79:275–282.
39 Zucht HD, Grabowsky J, Schrader M, Liepke C, Jurgens M, Schulz-Knappe P, Forssmann WG: Human beta-defensin-1: A urinary peptide present in variant molecular forms and its putative functional implication. Eur J Med Res 1998;3:315–323.
40 Circo R, Skerlavaj B, Gennaro R, Amoroso A, Zanetti M: Structural and functional characterization of hBD-1(Ser35), a peptide deduced from a DEFB1 polymorphism. Biochem Biophys Res Commun 2002;293:586–592.
41 Garcia JR, Krause A, Schulz S, Rodriguez-Jimenez FJ, Kluver E, Adermann K, Forssmann U, Frimpong-Boateng A, Bals R, Forssmann WG: Human beta-defensin 4: A novel inducible peptide with a specific salt-sensitive spectrum of antimicrobial activity. FASEB J 2001;15:1819–1821.
42 Ritonja A, Kopitar M, Jerala R, Turk V: Primary structure of a new cysteine proteinase inhibitor from pig leucocytes. FEBS Lett 1989;255:211–214.
43 Gudmundsson GH, Agerberth B, Odeberg J, Bergman T, Olsson B, Salcedo R: The human gene FALL39 and processing of the cathelin precursor to the antibacterial peptide LL-37 in granulocytes. Eur J Biochem 1996;238:325–332.
44 Agerberth B, Gunne H, Odeberg J, Kogner P, Boman HG, Gudmundsson GH: FALL-39, a putative human peptide antibiotic, is cysteine-free and expressed in bone marrow and testis. Proc Natl Acad Sci USA 1995;92:195–199.
45 Sorensen OE, Follin P, Johnsen AH, Calafat J, Tjabringa GS, Hiemstra PS, Borregaard N: Human cathelicidin, hCAP-18, is processed to the antimicrobial peptide LL-37 by extracellular cleavage with proteinase 3. Blood 2001;97:3951–3059.
46 Turner J, Cho Y, Dinh NN, Waring AJ, Lehrer RI: Activities of LL-37, a cathelin-associated antimicrobial peptide of human neutrophils. Antimicrob Agents Chemother 1998;42:2206–2214.
47 Wang Y, Agerberth B, Lothgren A, Almstedt A, Johansson J: Apolipoprotein A-I binds and inhibits the human antibacterial/cytotoxic peptide LL-37. J Biol Chem 1998;273:33115–33118.
48 Johansson J, Gudmundsson GH, Rottenberg ME, Berndt KD, Agerberth B: Conformation-dependent antibacterial activity of the naturally occurring human peptide LL-37. J Biol Chem 1998;273:3718–3724.
49 Zaiou M, Nizet V, Gallo RL: Antimicrobial and protease inhibitory functions of the human cathelicidin (hCAP18/LL-37) prosequence. J Invest Dermatol 2003;120:810–816.
50 De Y, Chen Q, Schmidt AP, Anderson GM, Wang JM, Wooters J, Oppenheim JJ, Chertov O: LL-37, the neutrophil granule- and epithelial cell-derived cathelicidin, utilizes formyl peptide receptor-like 1 (FPRL1) as a receptor to chemoattract human peripheral blood neutrophils, monocytes, and T cells. J Exp Med 2000;192:1069–1074.
51 Niyonsaba F, Iwabuchi K, Someya A, Hirata M, Matsuda H, Ogawa H, Nagaoka I: A cathelicidin family of human antibacterial peptide LL-37 induces mast cell chemotaxis. Immunology 2002; 106:20–26.
52 Scott MG, Davidson DJ, Gold MR, Bowdish D, Hancock RE: The human antimicrobial peptide LL-37 is a multifunctional modulator of innate immune responses. J Immunol 2002;169:3883–3891.
53 Frohm M, Agerberth B, Ahangari G, Stahle-Backdahl M, Liden S, Wigzell H, Gudmundsson GH: The expression of the gene coding for the antibacterial peptide LL-37 is induced in human keratinocytes during inflammatory disorders. J Biol Chem 1997;272:15258–15263.

54 Dorschner RA, Pestonjamasp VK, Tamakuwala S, Ohtake T, Rudisill J, Nizet V, Agerberth B, Gudmundsson GH, Gallo RL: Cutaneous injury induces the release of cathelicidin anti-microbial peptides active against group A Streptococcus. J Invest Dermatol 2001;117:91–97.
55 Murakami M, Ohtake T, Dorschner RA, Schittek B, Garbe C, Gallo RL: Cathelicidin anti-microbial peptide expression in sweat, an innate defense system for the skin. J Invest Dermatol 2002;119: 1090–1095.
56 Murakami M, Lopez-Garcia B, Braff M, Dorschner RA, Gallo RL: Postsecretory processing generates multiple cathelicidins for enhanced topical antimicrobial defense. J Immunol 2004;172: 3070–3077.
57 Marchini G, Lindow S, Brismar H, Stabi B, Berggren V, Ulfgren AK, Lonne-Rahm S, Agerberth B, Gudmundsson GH: The newborn infant is protected by an innate antimicrobial barrier: Peptide antibiotics are present in the skin and vernix caseosa. Br J Dermatol 2002;147:1127–1134.
58 Yoshio H, Tollin M, Gudmundsson GH, Lagercrantz H, Jornvall H, Marchini G, Agerberth B: Antimicrobial polypeptides of human vernix caseosa and amniotic fluid: Implications for newborn innate defense. Pediatr Res 2003;53:211–216.
59 Nizet V, Ohtake T, Lauth X, Trowbridge J, Rudisill J, Dorschner RA, Pestonjamasp V, Piraino J, Huttner K, Gallo RL: Innate antimicrobial peptide protects the skin from invasive bacterial infection. Nature 2001;414:454–457.
60 Fritz H: Human mucus proteinase inhibitor (human MPI). Human seminal inhibitor I (HUSI-I), antileukoprotease (ALP), secretory leukocyte protease inhibitor (SLPI). Biol Chem Hoppe Seyler 1988;369(suppl):79–82.
61 Wiedow O, Harder J, Bartels J, Streit V, Christophers E: Antileukoprotease in human skin: An antibiotic peptide constitutively produced by keratinocytes. Biochem Biophys Res Commun 1998;248:904–909.
62 Wingens M, van Bergen BH, Hiemstra PS, Meis JF, van Vlijmen-Willems IM, Zeeuwen PL, Mulder J, Kramps HA, van Ruissen F, Schalkwijk J: Induction of SLPI (ALP/HUSI-I) in epidermal keratinocytes. J Invest Dermatol 1998;111:996–1002.
63 Pfundt R, Wingens M, Bergers M, Zweers M, Frenken M, Schalkwijk J: TNF-alpha and serum induce SKALP/elafin gene expression in human keratinocytes by a p38 MAP kinase-dependent pathway. Arch Dermatol Res 2000;292:180–187.
64 Meyer-Hoffert U, Wichmann N, Schwichtenberg L, White PC, Wiedow O: Supernatants of *Pseudomonas aeruginosa* induce the *Pseudomonas*-specific antibiotic elafin in human keratinocytes. Exp Dermatol 2003;12:418–425.
65 Alkemade JA, Molhuizen HO, Ponec M, Kempenaar JA, Zeeuwen PL, de Jongh GJ, van Vlijmen-Willems IM, van Erp PE, van de Kerkhof PC, Schalkwijk J: SKALP/elafin is an inducible proteinase inhibitor in human epidermal keratinocytes. J Cell Sci 1994;107:2335–2342.
66 Simpson AJ, Maxwell AI, Govan JR, Haslett C, Sallenave JM: Elafin (elastase-specific inhibitor) has anti-microbial activity against gram-positive and gram-negative respiratory pathogens. FEBS Lett 1999;452:309–313.
67 Wiedow O, Schröder J-M, Gregory H, Young JA, Christophers E: Elafin: An elastase-specific inhibitor of human skin. Purification, characterization, and complete amino acid sequence. J Biol Chem 1990;265:14791–14795.
68 Simpson AJ, Wallace WA, Marsden ME, Govan JR, Porteous DJ, Haslett C, Sallenave JM: Adenoviral augmentation of elafin protects the lung against acute injury mediated by activated neutrophils and bacterial infection. J Immunol 2001;167:1778–1786.
69 Schittek B, Hipfel R, Sauer B, Bauer J, Kalbacher H, Stevanovic S, Schirle M, Schroeder K, Blin N, Meier F, Rassner G, Garbe C: Dermcidin: A novel human antibiotic peptide secreted by sweat glands. Nat Immunol 2001;2:1133–1137.
70 Zudaire E, Martinez A, Cuttitta F: Adrenomedullin and cancer. Regul Pept 2003;112:175–183.
71 Martinez A, Elsasser TH, Muro-Cacho C, Moody TW, Miller MJ, Macri CJ, Cuttitta F: Expression of adrenomedullin and its receptor in normal and malignant human skin: A potential pluripotent role in the integument. Endocrinology 1997;138:5597–5604.
72 Allaker RP, Zihni C, Kapas S: An investigation into the antimicrobial effects of adrenomedullin on members of the skin, oral, respiratory tract and gut microflora. FEMS Immunol Med Microbiol 1999;23:289–293.

73 Allaker RP, Kapas S: Adrenomedullin expression by gastric epithelial cells in response to infection. Clin Diagn Lab Immunol 2003;10:546–551.
74 Kapas S, Bansal A, Bhargava V, Maher R, Malli D, Hagi-Pavli E, Allaker RP: Adrenomedullin expression in pathogen-challenged oral epithelial cells. Peptides 2001;22:1485–1489.
75 Kjeldsen L, Johnsen AH, Sengelov H, Borregaard N: Isolation and primary structure of NGAL, a novel protein associated with human neutrophil gelatinase. J Biol Chem 1993;268:10425–10432.
76 Kjeldsen L, Bainton DF, Sengelov H, Borregaard N: Identification of neutrophil gelatinase-associated lipocalin as a novel matrix protein of specific granules in human neutrophils. Blood 1994;83:799–807.
77 Goetz DH, Holmes MA, Borregaard N, Bluhm ME, Raymond KN, Strong RK: The neutrophil lipocalin NGAL is a bacteriostatic agent that interferes with siderophore-mediated iron acquisition. Mol Cell 2002;10:1033–1043.
78 Weinberg ED: Human lactoferrin: A novel therapeutic with broad spectrum potential. J Pharm Pharmacol 2001;53:1303–1310.
79 Mallbris L, O'Brien KP, Hulthen A, Sandstedt B, Cowland JB, Borregaard N, Stahle-Backdahl M: Neutrophil gelatinase-associated lipocalin is a marker for dysregulated keratinocyte differentiation in human skin. Exp Dermatol 2002;11:584–591.
80 Harder J, Schröder J-M: RNase 7, a novel innate immune defense antimicrobial protein of healthy human skin. J Biol Chem 2002;277:46779–46784.
81 Rosenberg HF: Recombinant human eosinophil cationic protein. Ribonuclease activity is not essential for cytotoxicity. J Biol Chem 1995;270:7876–7881.
82 Chronnell CM, Ghali LR, Ali RS, Quinn AG, Holland DB, Bull JJ, Cunliffe WJ, McKay IA, Philpott MP, Muller-Rover S: Human beta defensin-1 and -2 expression in human pilosebaceous units: Upregulation in acne vulgaris lesions. J Invest Dermatol 2001;117:1120–1125.
83 Oono T, Huh WK, Shirafuji Y, Akiyama H, Iwatsuki K: Localization of human beta-defensin-2 and human neutrophil peptides in superficial folliculitis. Br J Dermatol 2003;148:188–191.
84 Conner K, Nern K, Rudisill J, O'Grady T, Gallo RL: The antimicrobial peptide LL-37 is expressed by keratinocytes in condyloma acuminatum and verruca vulgaris. J Am Acad Dermatol 2002;47: 347–350.
85 Ortega MR, Ganz T, Milner SM: Human beta defensin is absent in burn blister fluid. Burns 2000; 26:724–726.
86 Heilborn JD, Nilsson MF, Kratz G, Weber G, Sorensen O, Borregaard N, Stahle-Backdahl M: The cathelicidin anti-microbial peptide LL-37 is involved in re-epithelialization of human skin wounds and is lacking in chronic ulcer epithelium. J Invest Dermatol 2003;120:379–389.
87 Henseler T, Christophers E: Disease concomitance in psoriasis. J Am Acad Dermatol 1995;32: 982–986.
88 Ong PY, Ohtake T, Brandt C, Strickland I, Boguniewicz M, Ganz T, Gallo RL, Leung DY: Endogenous antimicrobial peptides and skin infections in atopic dermatitis. N Engl J Med 2002; 347:1151–1160.

Jürgen Harder, PhD
Clinical Research Unit, Department of Dermatology
University Hospital Schleswig-Holstein, Campus Kiel
Schittenhelmstrasse 7, DE–24105 Kiel (Germany)
Tel. +49 431 5971598, Fax +49 431 5971611, E-Mail jharder@dermatology.uni-kiel.de

Kabelitz D, Schröder J-M (eds): Mechanisms of Epithelial Defense.
Chem Immunol Allergy. Basel, Karger, 2005, vol 86, pp 42–54

Human Defensins in Crohn's Disease

A Molecular Link to Mucosal Barrier Dysfunction

Jan Wehkamp[a], *Klaus Fellermann*[b], *Eduard F. Stange*[b]

[a]Department of Microbiology and Immunology, University of California, Davis, Calif., USA; [b]Department of Internal Medicine I, Robert Bosch Hospital, Dr. Margarete Fischer Bosch Institute, Stuttgart, Germany

Abstract

Crohn's disease, a transmural inflammation of the gut, has been linked to good childhood hygiene, frequent use of antibiotics before diagnosis, adherent or invasive mucosal bacteria and a break in the tolerance of luminal bacteria. A decrease or lack of mucosal peptide antibiotics may play a central role in the etiopathogenesis of Crohn's disease. The dysregulated adaptive immune system may reflect only the primary break of the mucosal defence since the immune response is mostly directed against luminal bacteria. Crohn's disease patients with ileal involvement, as compared to controls and Crohn's disease patients without ileal disease, are characterized by a diminished expression of the ileal Paneth cell defensins. This decrease is even more pronounced in Crohn's disease patients with a mutation in the NOD2 gene, which is associated with Crohn's disease and ileal involvement. NOD2 is an intracellular peptidoglycan receptor and is expressed in Paneth cells. In contrast to ulcerative colitis, Crohn's disease of the colon is characterized by an impaired induction of human beta defensins 2 and 3. The hypothesis of an impaired mucosal antibacterial activity is also consistent with the benefit from antibiotic or probiotic treatment in certain disease states.

Introduction

Crohn's disease is a chronic disease of the intestine characterized by a transmural inflammation of the gut. Although the distal ileum is affected in about 70% of patients, the disease can be located anywhere from the oral cavity to the rectum. By comparison, ulcerative colitis is restricted to the colon and the inflammation is limited to the mucosa. Both diseases together constitute the two major chronic inflammatory bowel diseases affecting 1 in 500 individuals.

The principle treatment for both diseases is basically a more or less effective suppression of the inflammatory process with substantial side effects. Despite active research for many decades the etiology of Crohn's disease and ulcerative colitis is still enigmatic. Until recently, most of the research focused on a potential dysregulation of specific mucosal immunology. These investigations have elegantly described the mucosal immune cell populations and cytokine profiles associated with inflammatory bowel diseases, but have not succeeded in finding the etiological culprit. The alternative hypothesis of a primary defect in the mucosal barrier also has never been substantiated on the molecular level. In this article we outline a novel concept of how epidemiological, pathophysiological, genetic, molecular, clinical and pharmacological sets of data may be synthesized into a unifying hypothesis compatible with many features of this disease. A simplified illustration of this model is presented in figure 1.

Epidemiology: The Role of Hygiene

There is a clear-cut North to South gradient of inflammatory bowel disease incidences worldwide but also in Europe. In developing countries, infectious intestinal diseases represent the rule and idiopathic inflammatory bowel diseases, especially Crohn's disease, the rare exception. Since migration in many instances is associated with adaptation to the incidence rates in the host country, there is little doubt that environmental factors are involved. This is supported by the finding that good domestic hygiene in infancy has been shown to be a risk factor for Crohn's disease, but not for ulcerative colitis, even within the same country. Probably due to superior hygiene in infancy, *Helicobacter pylori* seroprevalence is substantially reduced in Crohn's disease (odds ratio 0.18) but not in ulcerative colitis. In addition, Crohn's disease occurs more often in members of small families than in those with many children. Since intra-familial transmission of common pathogens is frequent, the single child is particularly prone to be raised under more hygienic conditions with lower risk of acquiring gut infections. In addition, Crohn's disease patients are characterized by a frequent use of antibiotics [1]. Most likely, these various factors associated with the incidence of Crohn's disease serve as indicators of a rather clean environment leading to infrequent confrontations with pathogenic or nonpathogenic microorganisms. As a result, the intestinal innate immune system is probably not 'trained' to confront minor infections without recruiting the full array of specific immune function which acts only at the expense of relevant inflammation.

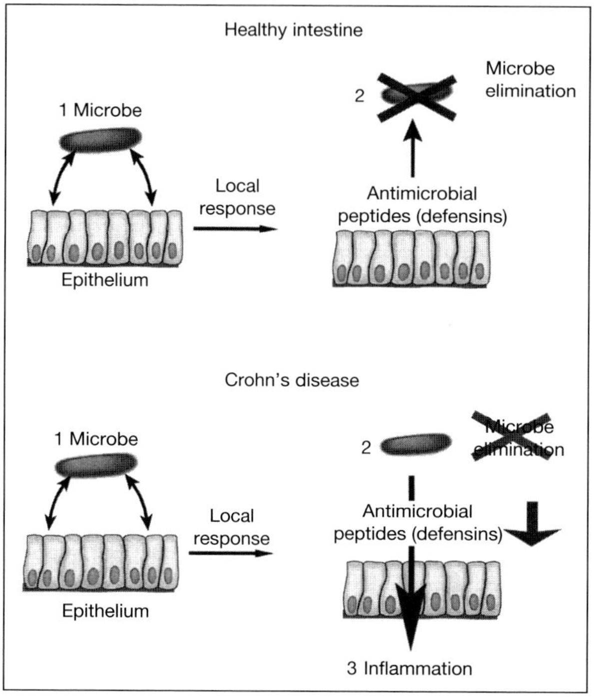

Fig. 1. Simplified model about the normal reaction (above) in the healthy intestine [modified from Huttner KM, Bevins CL: Pediatr Res 1999;45:785–794, with permission] as well as the defective antimicrobial barrier in the intestine of Crohn's disease patients. In the healthy gut, the microbes can not invade the mucosa because of an effective antimicrobial barrier (defensins). In Crohn's disease (below), especially in patients with ileal affection, this antimicrobial barrier is disturbed and bacteria can invade the mucosa. According to this hypothesis, a bacterial invasion due to a defensin deficiency is the primary reason for the secondary inflammation.

Another important aspect is the apparently frequent association of a recent intestinal infection with the first appearance of Crohn's disease, and the prevalence of superinfection in pre-established inflammatory bowel disease. Although these relationships are not fully understood, the classical interpretation made is that an infection in some way triggers a relapse of the idiopathic bowel disease by breaking mucosal tolerance. Despite their self-limited character, these infections may initiate a cascade of inflammatory events leading to chronic relapsing disease in genetically susceptible hosts (the 'hit and run' hypothesis). Alternatively, the host with inflammatory bowel disease may be more likely to contract an intestinal infection because of a defective innate defence system.

Pathophysiology: The Role of Luminal and Mucosal Bacteria

It has always been an intriguing hypothesis that inflammatory bowel diseases are caused by a specific, hitherto unrecognized infection. For example, *Mycobacterium paratuberculosis* has been considered by various groups to cause not only Johne's disease in cattle but also Crohn's disease in humans. The debate has been ongoing for many years and is beyond the scope of the present considerations. It should be noted, however, that even very recently, using novel techniques like granuloma isolation with laser capture microdissection, many more Crohn's samples were shown to be positive for *Mycobacteria* than controls. Thus, although *Mycobacteria* are far from proven to be causative agents, it is apparent that the mucosa in Crohn's disease frequently harbors unusual and potentially pathogenic bacteria. In some instances *Listeria* has been isolated or specific mucosal adherent *Escherichia coli* [2]. Interestingly, there is a tremendous increase in the mucosal-associated bacterial counts in the neoterminal ileum after ileocecal resection for Crohn's disease, and this colonization may be related to postoperative relapse. Measles infection in Crohn's disease is also a very controversial issue but it adds to the list of transmissible agents recovered from Crohn's mucosa [3]. Taken together, these findings indicate that Crohn's mucosa is often the target of various infections but the proof that the disease is caused by these agents is missing. Most importantly, the immune response in the gut mucosa is not specific for any of these suspicious agents, but rather unspecific to a multitude of organisms.

It has only recently been appreciated that the mucosal immune response in inflammatory bowel disease is directed towards a multitude of common luminal bacteria. The most convincing evidence for a break in mucosal tolerance in intestinal inflammation stems from the observation that knockout mice lacking one of several relevant genes, including interleukins 2 or 10, develop experimental colitis only when raised in contaminated but not in sterile conditions [4]. This fits well with the consistent finding of a break in mucosal tolerance towards various luminal bacteria in inflammatory bowel diseases [5]. It may be concluded that these diseases are not autoimmune-mediated in the strict sense, i.e. due to reactivity against autologous tissues, but only in a more general sense, i.e. due to immune responses towards commensal bacteria. The permeable mucosal barrier may also explain the development of anti-*Saccharomyces cerevisiae* antibodies especially in familial Crohn's disease as well as antibodies to various other microbes, including *E. coli*.

The most surprising finding in this regard is the demonstration by Swidsinski et al. [6] that the mucosa in inflammatory bowel diseases is heavily contaminated by adherent and sometimes invading bacteria entering from the lumen. In contrast, normal mucosa is virtually sterile when washed a few times in saline. Indirectly, a change in the expression or function of this chemical defence may indeed explain the changes in bacterial flora in inflammatory bowel diseases reviewed by Linskens et al. [7].

These findings are difficult to reconcile with an immunological dysregulation as the sole basis for intestinal inflammation in these diseases. Rather, it is conceivable that there may be a primary defect in the chemical barrier of intestinal antibiotic defensins, which protect the normal mucosa extremely efficiently against adherent or entering microbes. We believe that a thorough understanding of these functionally relevant peptides is paramount to understanding the true pathogenesis of inflammatory bowel disease.

Defensin Expression and Regulation in the Healthy Intestinal Tract

Many important studies in the field of defensins have focused on the human and non-vertebrate skin as another border of the body exposed to a multitude of bacteria. These studies resulted in the isolation of various peptides exhibiting potent antibiotic activity towards both gram-positive and gram-negative bacteria, as well as enveloped viruses and fungi [8–10]. A similar system of antibiotic peptides is apparently synthesized and secreted by the intestinal mucosa as part of innate immunity, but has received little attention at least in the field of clinical gastroenterology. Six α-defensins and 4 β-defensins have been identified in humans so far. The α-defensins comprise human neutrophil peptide 1–4, abundant in granulocytes, and human defensin 5 and 6, synthesized in Paneth cells [11]. The β-defensins are of epithelial origin and abundant in skin, urogenital tract, intestine and lung. The concept of a certain defensin exclusively formed by specialized tissues or cells has to be revised as inflammation induces epithelial expression of human neutrophil peptides and β-defensins in monocytes and lymphocytes [12].

Defensins can be divided into constitutive forms, e.g. HBD-1 with its widespread stable distribution [13] and inducible peptides like HBD-2 [8]. The mechanisms of activation are still under investigation. Induction by cytokines, such as IL-1β and TNF-α, has been shown in addition to a direct response to bacterial components, such as lipopolysaccharides and lipoproteins. Possible signalling pathways involve Toll-like receptors (TLRs), especially TLR2 and TLR4, eventually leading to NFκB-mediated activation of transcription. NOD2/CARD15 as an intracellular peptidoglycan receptor inducing NFκB activation [14], which in turn is known to trigger HBD-2 transcription. Interestingly, this NFκB response is impaired in the NOD2 insertion mutation associated with Crohn's disease [15, 16], suggesting a diminished innate

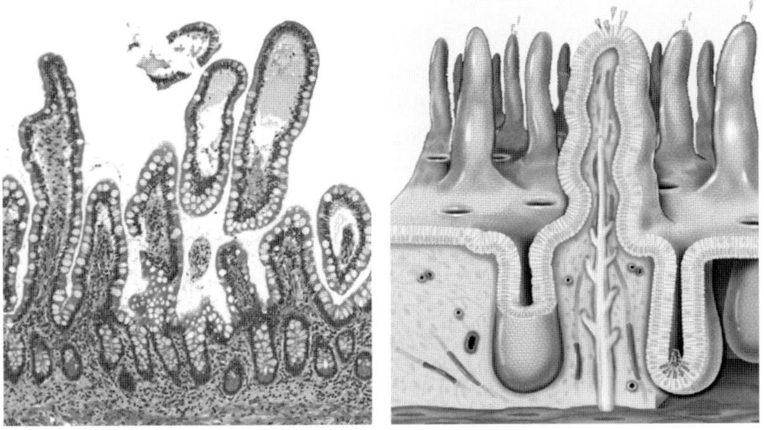

a *b*

Fig. 2. Hematoxylin eosin (HE) staining (***a***) as well as schematic demonstration (***b***) of the small intestine. Paneth cells are expressed at the base of the crypts. Paneth cells are the only source of α-defensins HD-5 and HD-6 which are the major antibacterial factors in the ileum. In mice, Paneth cell defensins contribute to 70% of antibiotic activity [24]. From Ganz T: Paneth cells: Guardians of the gut cell hatchery, Nat Immunol 2000;1:99–100, with permission.

response to bacterial components. Human defensin 5 is released as a propeptide from Paneth cells and activated by trypsinogen in the lumen of the intestinal crypts [17]. Figure 2 shows HE staining as well as a graphical illustration of ileal Paneth cells. The functional significance in bacterial infection has recently been shown in HD-5 transgenic mice which are protected from lethal *Salmonella* infection [18]. On the other hand, mice deficient in the metalloproteinase matrilysin fail to process defensins efficiently and exhibit higher bacterial counts [19].

Defensins and Inflammatory Bowel Diseases

Some defensins appear to be induced in both Crohn's disease and ulcerative colitis. Human neutrophil peptides 1–3, as well as lysozyme, are expressed in surface enterocytes of mucosa with active IBD, but surprisingly not in controls [20]. HD-5 is stored in a precursor form in normal Paneth cells and is expressed by metaplastic colonic Paneth cells. Metaplastic Paneth cells express both α-defensins HD-5 and HD-6 in the colonic mucosa of inflammatory bowel disease patients [21, 22]. A very recent study showed a decreased ileal Paneth cell defensin HD-5 and HD-6 expression in patients with ileal disease as compared to controls and colonic non-ileal Crohn's disease. This difference was even more pronounced in patients with a NOD2 mutation (see below) [23]. The functional significance of HD-5 [18] as well as the fact that mouse Paneth cell defensins contribute to about 70% of the antibiotic activity in the ileum [24] illustrate the potential importance of these findings.

As opposed to the low α-defensin expression in the ileum, the β-defensins show a conspicuous difference between Crohn's disease and ulcerative colitis in the colon. It has been suggested that HBD-1 is constitutively expressed in the intestinal epithelium [25] and qualitative investigations indeed showed constitutive expression in normal tissue and inflammatory bowel disease mucosa [26]. With the quantitative approach, a paradoxical decrease of HBD-1 was found in inflamed mucosa of both Crohn's disease and ulcerative colitis patients, as shown in figure 3 [27]. However, it remains to be shown if such a decrease actually translates into a diminished mucosal antibacterial activity.

The inducible HBD-2, which has been described originally in skin [8], is also expressed in the colon during inflammation [25], particularly in ulcerative colitis [26]. It has now been shown by various independent groups that HBD-2 is highly induced in inflamed mucosa of ulcerative colitis patients. As compared to ulcerative colitis, this induction is lacking in Crohn's disease [26–28]. Most likely, there is a lack of β-defensin induction in Crohn's disease contributing to a defective antimicrobial barrier or, alternatively, there is an excessive induction in ulcerative colitis. The third defensin studied was HBD-3 which was reported by Harder et al. [9] as a novel inducible β-defensin in skin. Another group described HBD-3 based on genomic analysis [29]. Our study was the first to describe HBD-3 in the human colon. Although HBD-3 was also slightly induced in inflamed Crohn's mucosa, its expression was preferentially enhanced in inflamed and non-inflamed ulcerative colitis [27]. A deficiency in the antimicrobial defense systems of defensins may be a reasonable and plausible explanation for the break of the antibacterial barrier function in inflammatory bowel diseases.

In conclusion, the most fascinating finding in this context to date is the association of ileal disease with low Paneth cell defensin expression. On the other hand, the decrease of HBD-1 in both inflammatory bowel diseases and the lack of induction of both inducible β-defensins HBD-2 and HBD-3 in Crohn's disease could contribute to a deficient mucosal barrier function in the colon. This may in part be compensated by the expression of Paneth cell α-defensins in the colon. A lack in the innate defense system of antimicrobial peptides may lead to a permanent but slow bacterial invasion triggering the inflammatory process but further direct studies on antimicrobial peptide activity in inflammatory bowel disease mucosa are required to validate this hypothesis.

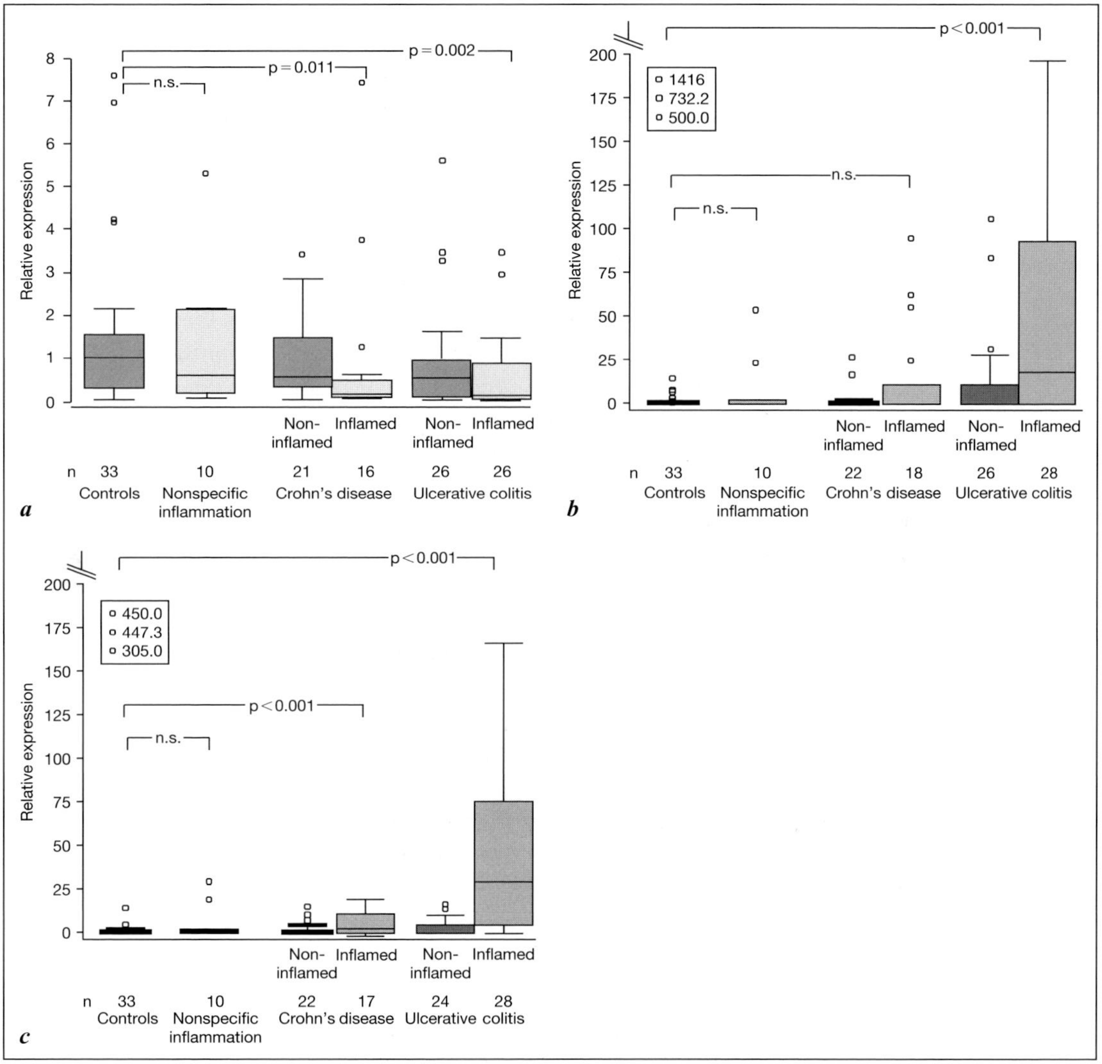

Fig. 3. Colonic expression of HBD-1 (***a***), HBD-2 (***b***) and HBD-3 (***c***) in healthy controls and patients with unspecific colitis, ulcerativce colitis and Crohn's disease. Boxes represent the interquartile range (25th and 75th percentiles) of values with the median given as horizontal line. In Crohn's disease and ulcerative colitis, the light bars denote specimens from macroscopically inflamed areas, while dark bars denote specimens from uninvolved mucosa, respectively. n.s. = Not significant. Figures from Wehkamp et al. [27], with permission.

NOD2, A Peptidoglycan Receptor and Defensin Expression

Although the etiology of Crohn's disease is still enigmatic, the recent finding that about a third of Crohn's disease patients have a loss of function mutation in the putative intracellular peptidoglycan receptor NOD2 represents a major advance. The pathophysiology of NOD2 in Crohn's disease was proposed to link to immunological dysregulation in monocytes. Alternatively, intestinal epithelial cells and Paneth cells [30], which have also been demonstrated to express this putative receptor, might be compromised in their antibacterial response. It has been demonstrated that NOD2 mutated epithelial cells display a disturbed response against *Salmonella* [31], supporting the hypothesis of a defensin deficiency. A new study from our laboratory shows that a mutation in the NOD2 gene leads to a disturbed expression of antimicrobial Paneth cell defensins in the colon and ileum. In the case of the NOD2 mutation, the colonic mucosa seems incapable of expressing HD-5 and HD-6 [23]. In the ileum, where the expression levels of Paneth cell defensins are much higher, expression of HD-5 and HD-6 is decreased in the case of the mutation. Interestingly, the question of whether or not a Crohn's disease patient has an ileal involvement seems to be determined by the expression of Paneth cell defensins. Even in patients with normal NOD2 genotype, the expression of HD-5 is especially low in case of ileal involvement as compared to controls or colonic-only involvement of Crohn's disease [23]. This disturbed Paneth cell defensin expression might explain the ileal preference in NOD2-mutated Crohn's disease patients.

Toll-Like Receptors and Their Expression in Inflammatory Bowel Diseases

Intestinal epithelial cells express various pattern recognition receptors recognizing microbial 'pathogen-associated molecular patterns' as 'non-self' to rapidly initiate innate immune responses of survival and to activate defence strategies against luminal pathogens. This system of several functional TLRs appears to be a key regulator of the innate response system. Different TLRs are responding to different pathogens and bacterial components, including lipopolysaccharide, flagellin and others. In active IBD, the expression of TLR3 and TLR4 is differentially modulated in the intestinal epithelium. TLR3 is significantly downregulated in active Crohn's disease (CD) but not in ulcerative colitis (UC), while TLR4 is upregulated in both UC and CD. TLR2 and TLR5 expression remained unchanged in IBD [32]. These data suggest that IBD may be associated with distinctive changes in selective TLR expression in the intestinal epithelium, implying that alterations in the innate response system may contribute to the pathogenesis of these disorders. The recent report of functionally relevant polymorphisms of TLRs may well be related to defensin expression.

Therapy: The Role of Antibiotics and Probiotics

If a deficiency of these endogenous antibiotics was triggering relapse one would expect exogenous antibiotics to be an efficacious treatment option. Indeed, antibiotics appear to have a limited effect in Crohn's disease and similarly, probiotics in ulcerative colitis. In Crohn's disease, the exogenous antibiotics may compensate for the deficient endogenous antibiotic response to infection or commensal bacterial invasion. In ulcerative colitis, the pattern is different with low basal activity but normal induction during inflammation. Therefore, antibiotics may not work and the benefit of probiotics may be due to the induction of β-defensins as demonstrated recently in vitro. In contrast to more than 40 tested *E. coli*, the probiotic Nissle 1917 as well as other probiotic strains potently upregulated HBD-2 expression in intestinal epithelial cells [33]. The same strain has been shown in three randomized controlled trials to maintain remission in ulcerative colitis.

Concluding Remarks

Seventy years after Crohn's description of the disease named after him it becomes apparent that Crohn's disease is not a disease but a syndrome. It is not surprising that the diverse facets of genetic predisposition, where only a minority of patients display a defective NOD2 gene, modified by environmental factors like childhood hygiene and others, may lead to very different forms of disease with respect to localization, natural course and therapeutic response. Although in no way perfect, the present hypothesis appears to be plausible for the reasons presented above, but particularly since the multitude of defensins, other antibiotic peptides, and related transcription factors or transporters leaves enough room for clinical diversity among patients. The link between mutations of the NOD2 gene in Crohn's disease patients related to a diminished defensin expression gives further evidence for our hypothesis that Crohn's disease may be a defensin deficiency syndrome [34].

Acknowledgements

Helpful discussions and joint investigations with J.-M. Schröder, J. Harder and C. Bevins are gratefully acknowledged. The work was supported generously by the Robert Bosch Foundation, Stuttgart, Germany.

References

1 Card T, Logan RF, Rodrigues LC, Wheeler JG: Antibiotic use and the development of Crohn's disease. Gut 2004;53:246–250.
2 Darfeuille-Michaud A, Neut C, Barnich N, Lederman E, di Martino P, Desreumaux P, Gambiez L, Joly B, Cortot A, Colombel JF: Presence of adherent *Escherichia coli* strains in ileal mucosa of patients with Crohn's disease. Gastroenterology 1998;115:1405–1413.
3 Wakefield AJ, Ekbom A, Dhillon AP, Pittilo RM, Pounder RE: Crohn's disease: Pathogenesis and persistent measles virus infection. Gastroenterology 1995;108:911–916.
4 Sellon RK, Tonkonogy S, Schultz M, Dieleman LA, Grenther W, Balish E, Rennick DM, Sartor RB: Resident enteric bacteria are necessary for development of spontaneous colitis and immune system activation in interleukin-10-deficient mice. Infect Immun 1998;66: 5224–5231.
5 Duchmann R, May E, Heike M, Knolle P, Neurath M, Meyer zum Büschenfelde KH: T cell specifity and cross reactivity towards enterobacteria, *Bacteroides*, *Bifidobacterium*, and antigens from resident intestinal flora in humans. Gut 1999;44:812–818.
6 Swidsinski A, Ladhoff A, Pernthaler A, Swidsinski S, Loening-Baucke V, Ortner M, Weber J, Hoffmann U, Schreiber S, Dietel M, Lochs H: Mucosal flora in inflammatory bowel disease. Gastroenterology 2002;122:44–54.
7 Linskens RK, Huijsdens XW, Savelkoul PH, Vandenbroucke-Grauls CM, Meuwissen SG: The bacterial flora in inflammatory bowel disease: Current insights in pathogenesis and the influence of antibiotics and probiotics. Scand J Gastroenterol Suppl 2001;234:29–40.
8 Harder J, Bartels J, Christophers E, Schröder J-M: A peptide antibiotic from human skin. Nature 1997;387:861.
9 Harder J, Bartels J, Christophers E, Schröder J-M: Isolation and characterization of human9β-defensin-3, a novel human inducible peptide antibiotic. J Biol Chem 2001;276: 5707–5713.
10 Zasloff M: Antibiotic peptides as mediators of innate immunity. Curr Opin Immunol 1992;4:3–7.
11 Jones DE, Bevins CL: Paneth cells of the human small intestine express an antimicrobial peptide gene. J Biol Chem 1992;267:23216–23225.
12 Duits LA, Ravensbergen B, Rademaker M, Hiemstra PS, Nibbering PH: Expression of beta-defensin 1 and 2 mRNA by human monocytes, macrophages and dendritic cells. Immunology 2002;106:517–525.
13 Zhao C, Wang I, Lehrer RI: Widespread expression of beta-defensin hBD-1 in human secretory glands and epithelial cells. FEBS Lett 1996;396:319–322.
14 Ogura Y, Bonen DK, Inohara N, Nicolae DL, Chen FF, Ramos R, Britton H, Moran T, Karalluskas R, Duerr RH, Achkar JP, Brant SR, Bayless TM, Kirschner BS, Hanauer SB, Nunez G, Cho JH: A frameshift mutation in NOD2 associated with susceptibility to Crohn's disease. Nature 2001;411: 603–606.
15 Hugot JP, Chamaillard C, Zouali H, Lesage S, Cezard JP, Belaiche J, Almer S, Tysk C, O'Morain CA, Gassull M, Binder V, Finkel Y, Cortot A, Modigliani R, Laurent-Puig P, Gower-Rousseau C, Macry J, Colombel JF, Sahbatou M, Thomas G: Association of NOD2 leucine-rich repeat variants with susceptibility to Crohn's disease. Nature 2001;411:599–603.
16 Ogura Y, Inohara N, Benito A, Chen FF, Yamaoka S, Nunez G: Nod2, a Nod1/Apaf-1 family member that is restricted to monocytes and activates NF-kappaB. J Biol Chem 2001;276: 4812–4818.

17 Ghosh D, Porter E, Shen B, Lee SK, Wilk D, Drazba J, Yadav SP, Crabb JW, Ganz T, Bevins CL: Paneth cell trypsin is the processing enzyme for human defensin-5. Nat Immunol 2002;3: 583–590.

18 Salzman NH, Ghosh D, Huttner KM, Paterson Y, Bevins CL: Protection against enteric salmonellosis in transgenic mice expressing a human intestinal defensin. Nature 2003;422: 522–526.

19 Wilson CL, Ouellette AJ, Satchell DP, Ayabe T, Lopez-Boado YS, Stratman JL, Hultgren SJ, Matrisian LM, Parks WC: Regulation of intestinal α-defensin activation by the metalloproteinase matrilysin in innate host defense. Science 1999;286:113–117.

20 Cunliffe RN, Kamal M, Rose FR, James PD, Mahida YR: Expression of antimicrobial neutrophil defensins in epithelial cells of active inflammatory bowel disease mucosa. J Clin Pathol 2002;55:298–304.

21 Cunliffe RN, Rose FR, Keyte J, Abberley L, Chan WC, Mahida YR: Human defensin 5 is stored in precursor form in normal Paneth cells and is expressed by some villous epithelial cells and by metaplastic Paneth cells in the colon in inflammatory bowel disease. Gut 2001;48: 176–185.

22 Wehkamp J, Schwind B, Herrlinger KR, Baxmann S, Schmidt K, Duchrow M, Wohlschlaeger C, Feller AC, Stange EF, Fellermann K: Innate immunity and colonic inflammation: enhanced expression of epithelial alpha-defensins. Dig Dis Sci 2002;47:1349–1355.

23 Wehkamp J, Harder J, Weichenthal M, Schwab M, Schaeffeler E, Schlee M, Stallmach A, Noack F, Fritz P, Schröder J-M, Bevins CL, Fellermann K, Stange EF: NOD2 (CARD15) mutations in Crohn's disease are associated with diminished mucosal alpha-defensin expression. Gut 2004;53:1658–1664.

24 Ayabe T, Satchell DP, Wilson CL, Parks WC, Selsted ME, Ouellette AJ: Secretion of microbicidal alpha-defensins by intestinal Paneth cell response to bacteria. Nat Immunol 2000;1: 113–118.

25 O'Neil DA, Porter EM, Elewaut D, Anderson GM, Eckmann L, Ganz T, Kagnoff MF: Expression and regulation of the human β-defensins hBD-1 and hBD-2 in intestinal epithelium. J Immunol 1999;163:6718–6724.

26 Wehkamp J, Fellermann K, Herrlinger KR, Baxmann S, Schmidt K, Schwind B, Duchrow M, Wohlschlager C, Feller AC, Stange EF: Human beta-defensin 2 but not beta-defensin 1 is expressed preferentially in colonic mucosa of inflammatory bowel disease. Eur J Gastroenterol Hepatol. 2002;14:745–752.

27 Wehkamp J, Harder J, Weichenthal M, Mueller O, Herrlinger KR, Fellermann K, Schroeder JM, Stange EF: Inducible and constitutive beta-defensins are differentially expressed in Crohn's disease and ulcerative colitis. Inflamm Bowel Dis 2003;9:215–223.

28 Fahlgren A, Hammarstrom S, Danielsson A, Hammarstrom ML: Increased expression of antimicrobial peptides and lysozyme in colonic epithelial cells of patients with ulcerative colitis. Clin Exp Immunol 2003;131:90–101.

29 Garcia JR, Jaumann F, Schulz S, Krause A, Rodriguez-Jimenez J, Forssmann U, Adermann K, Kluver E, Vogelmeier C, Becker D, Hedrich R, Forssmann WG, Bals R: Identification of a novel, multifunctional beta-defensin (human beta-defensin 3) with specific antimicrobial activity. Its interaction with plasma membranes of Xenopus oocytes and the induction of macrophage chemoattraction. Cell Tissue Res 2001;306:257–264.

30 Lala S, Ogura Y, Osborne C, Hor SY, Bromfield A, Davies S, Ogunbiyi O, Nunez G, Keshav S: Crohn's disease and the NOD2 gene: A role for Paneth cells. Gastroenterology 2003;125: 47–57.

31 Hisamatsu T, Suzuki M, Reinecker HC, Nadeau WJ, McCormick BA, Podolsky DK: CARD15/NOD2 functions as an antibacterial factor in human intestinal epithelial cells. Gastroenterology 2003;124:993–1000.

32 Cario E, Podolsky DK: Differential alteration in intestinal epithelial cell expression of toll-like receptor 3 (TLR3) and TLR4 in inflammatory bowel disease. Infect Immun 2000;68: 7010–7017.

33 Wehkamp J, Harder J, Wehkamp K, Wehkamp v.Meissner B, Schlee M, Enders C, Sonnenborn U, Nuding S, Fellermann K, Schroeder JM, Stange EF: NF-kB and AP-1 mediated induction of human beta defensin-2 in intestinal epithelial cells by *E. coli* Nissle 1917: A novel effect of a probiotic bacterium. Infect Immun 2004;72:5750–5758.

34 Fellermann K, Wehkamp J, Herrlinger KR, Stange EF: Crohn's disease: A defensin deficiency syndrome? Eur J Gastroenterol Hepatol 2003;15:627–634.

Prof. Eduard F. Stange, MD
Department of Internal Medicine I, Robert Bosch Hospital
Auerbachstrasse 110, DE–70376 Stuttgart (Germany)
Tel. +49 711 8101 3406, Fax +49 711 8101 3793, E-Mail eduard.stange@rbk.de

Kabelitz D, Schröder J-M (eds): Mechanisms of Epithelial Defense.
Chem Immunol Allergy. Basel, Karger, 2005, vol 86, pp 55–71

Antimicrobial Peptides in Lung Inflammation

Christoph Beisswenger, Robert Bals

Department of Internal Medicine, Division of Pulmonology, Hospital of the University of Marburg, Philipps-Universtät Marburg, Marburg, Germany

Abstract

Antimicrobial peptides (AMPs) are expressed in the respiratory tract and act as effector substances of the innate immune system. A variety of cells synthesize and secrete AMPs including epithelial and professional host defense cells such as neutrophils, macrophages, and NK cells. In the human lung, β-defensins originate from epithelial cells, macrophages and lymphocytes. α-defensins are synthesized by neutrophils. LL-37/hCAP-18 is produced by epithelial cells, neutrophils, lymphocytes, and macrophages. AMPs act as endogenous antibiotics by direct destruction of microorganisms. Recently, it became clear that AMPs bind to cellular receptors and activate a variety of cell types such as airway epithelial cells, endothelial cells, mast cells, macrophages, dendritic cells, and neutophils amongst others. Concentrations of AMPs in lung secretions are altered in several pulmonary diseases. This chapter describes the basic and applied biology of AMPs in the human lung and their potential role in pulmonary disease.

Introduction

Findings in 1996 that airway epithelia from cystic fibrosis patients failed to kill bacteria led to increased interest in antimicrobial factors in airway secretions and the role of antimicrobial peptides (AMPs) in the lung [1]. AMPs act as endogenous antibiotics and are involved in a first line host defense. Recent insight into the basic biology of AMPs showed that these molecules have various functions in host defense, inflammation, and tissue regeneration and are likely to be involved in the pathogenesis of several diseases. It is the aim of this chapter to summarize the current knowledge about AMPs in the respiratory tract and their role in pulmonary disease.

AMPs Are Expressed in the Respiratory Tract

Host Defense in the Airways

The respiratory tract is shielded by a multi-component host defense system that involves structural, physical and functional mechanisms [2]. Different cell types of the innate and adaptive immune system have been described in the lungs: alveolar macrophages, neutrophils, eosinophils, dendritic cells, mast cells, NK cells, and lymphocytes. Also, parenchymal cells such as airway and alveolar epithelial cells contribute to host defense [3]. Secretions of the airways contain proteins and peptides that directly kill pathogens or modulate the inflammatory response. Classical components of the airway surface fluid (ASF) that have antimicrobial activity are lysozyme, lactoferrin, secretory phospholipase A2, and secretory leukocyte protease inhibitor (SLPI). Other substances, such as complement, surfactant proteins, Clara-cell proteins (CC10, CCSP), and proteins from the PLUNC family [4] likely contribute to host defense. Cationic polypeptides have been identified as an essential part of the antimicrobial activity of human airway fluid. AMPs have several cellular sources: epithelial cells of the airways, including serous gland cells, and host defense cells such as neutrophils, macrophages and NK cells (fig. 1).

AMPs in the Human Lung

AMPs can be grouped according to size, conformational structure, or predominant amino acids; however, the diversity of the molecules is so great that it is difficult to categorize them in a generally accepted classification. The principal families found in the respiratory tract are the defensins and the cathelicidins. General features of these families are described in earlier chapters.

Defensins. Human neutrophil peptides 1–4 (HNP 1–4) are localized in azurophilic granules of neutrophils where they represent the principal protein and contribute to the oxygen-independent killing of phagocytozed microorganisms [5, 6]. The two other α-defensins in humans, i.e. defensins 5 and 6 (HD 5–6), are primarily found in Paneth cells of the small intestine. HNP 1–4 are present in airway secretions, originating from neutrophils that invaded the airway or alveolar lumen. The first human β-defensin, called human β-defensin 1 (hBD-1), was isolated from large volumes of hemofiltrate [7] and is expressed constitutively in epithelial cells of the urinary and respiratory tracts [8–10]. Human β-defensin 2 (hBD-2) was isolated from psoriatic skin using an affinity chromatography procedure with columns coated with components of *Escherichia coli* [11], and from airway epithelium by a strategy based on database screening [12]. hBD-2 is expressed in epithelia of the respiratory tract and other body surfaces [12, 13]. Additionally, hBD-1 and hBD-2 are expressed by monocytes,

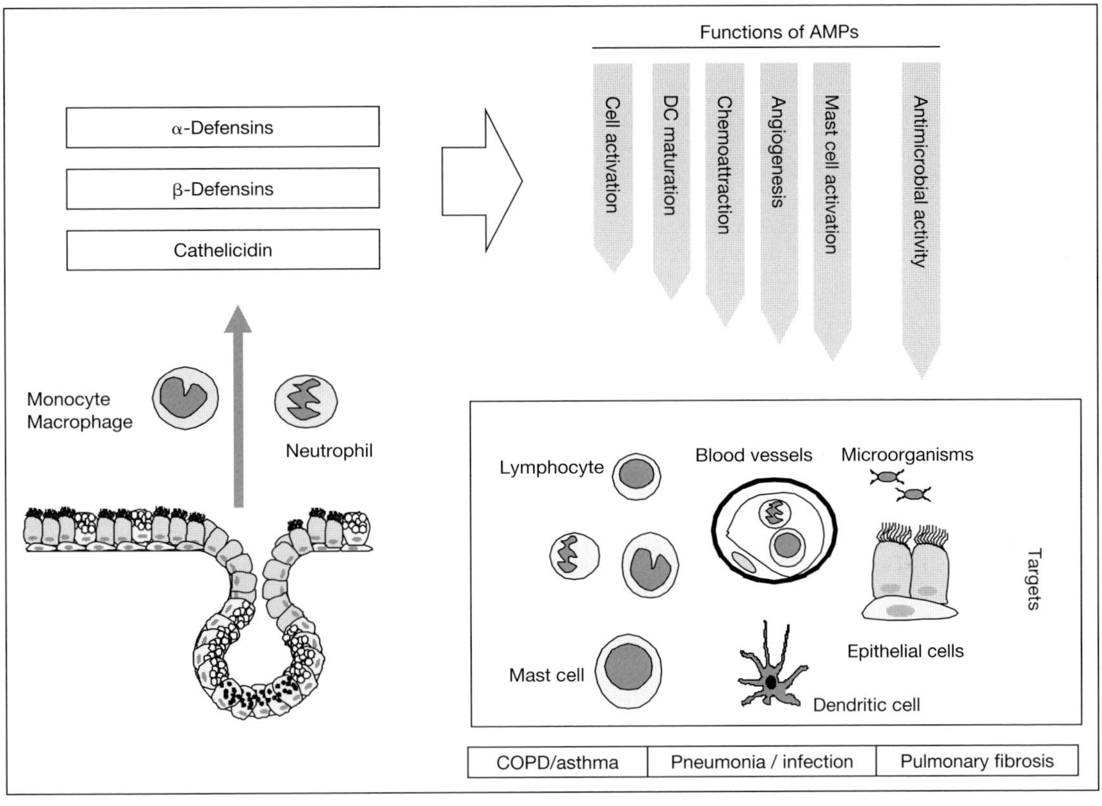

Fig. 1. Sources and functions of AMPs in the respiratory tract. Besides their direct antimicrobial function, AMPs act as immune mediators by binding to specific receptors on target cells. It is likely that these functions play a role in the listed pulmonary diseases.

macrophages, and dendritic cells [14]. Both peptides have been detected in airway secretions at μg/ml concentrations [12, 13]. hBD-3 was identified in parallel using bioscreening and computational approaches [15–17], while hBD-4 was identified solely by searches of genomic databases [18]. Recently, up to 28 new human and 43 new mouse beta-defensin genes in five syntenic chromosomal regions have been identified by computer-based screening of the human and murine genomes [19]. A novel class of defensins was recently isolated from rhesus monkey neutrophils and named theta-defensins according to their circular molecular structure [20].

Cathelicidins. Peptide antibiotics of the cathelicidin family contain a highly conserved signal sequence and pro-region (termed 'cathelin' = cathepsin L inhibitor) but show substantial heterogeneity in the C-terminal domain that encodes the mature peptide, which can range in size from 12 to 80 or more amino acids [21]. The only human cathelicidin, LL-37/hCAP-18, was isolated from human bone marrow [22, 23]. The peptide is expressed in myeloid cells where it resides in granules. In the airways, the peptide is produced by epithelial cells, macrophages, lymphocytes, and neutrophils and secreted into the ASF [24, 25]. LL-37 has been detected in tissue culture supernatants of respiratory epithelial cells as well as in lung washings from patients [24, 26]. At this time no details are known about the processing of LL-37/hCAP-18 in the pulmonary tract. In neutrophils, where LL-37/hCAP-18 is localized to specific granules, the peptide is stored in its pro-peptide form and cleaved after secretion by the activity of protease 3 [27]. hCAP-18 is processed in seminal plasma to generate the novel AMP ALL-38 by the protease gastricsin [28]. The cathelin-like prosequence has significant antimicrobial activity and acts as an antiprotease [29]. Further structural and functional diversity may be generated by postsecretory processing [30].

Granulysin [31], histatins, [32], and calcitermin [33] are other human AMPs. Their role in lung biology is unclear.

Regulation of the AMPs in the Lung

The expression of AMPs of the lung is tightly regulated. A first dimension of regulation concerns the cell type specificity of expression. A second dimension concerns the timeline of regulation after a specific stimulus. hBD-1 is constitutively expressed in various in vitro models [13], whereas increased concentrations of the peptide can be found in patient studies [34]. Expression of hBD-2, hBD-3, and hBD-4 is induced by proinflammatory stimuli or microorganisms [11, 12, 16–18, 35, 36]. hBD-2 signaling pathways involve NF-κB [37] and mitogen-activated protein kinases [38] including Src-dependent Raf-MEK1/2-ERK [39]. The promoter of hBD-2 has binding sites for NF-κB, AP-1, NF-IL-6, and STATs [37, 40]. In epithelial cells the transcription factor MEF acts as a transactivator of hBD-2, which functions through a mechanism independent of NF-κB [41]. Furthermore it is known that AP-1 and intracellular calcium are also involved in hBD-2 expression in airway epithelial cells [42]. Human airway epithelial cells grown at air-liquid interface respond to bacterial lipopeptide in a Toll-like receptor (TLR)-2 dependent manner with induction of mRNA and protein of hBD-2 [40, 43, 44]. hBD-2 is also up-regulated by LPS through a CD14-dependent mechanism [45]. Furthermore, expression of hBD-2 in airway epithelial cells appears to be driven by cytokines secreted by activated macrophages [46]. In contrast, the murine β-defensin mBD-2 also seems to act as endogenous danger signal and to bind to Toll-like receptors [47]. Neutrophil elastase induces hBD-2 expression in airway epithelial cells [48]. The expression of LL-37 is induced by inflammatory or infectious stimuli [23, 49], however, this has not been shown in vitro for airway epithelial cells. Cell differentiation appears to be a key determinant of LL-37/hCAP-18 expression by human colon epithelium [50]. The expression of defensins and LL-37 is downregulated by chronic inflammation, including a Th2 biased inflammation [51, 52]. In cystic fibrosis lung disease defensins are relatively suppressed [34]. Gastrointestinal *Shigella dysenteriae* infections in humans are associated with reduced expression of LL-37 in epithelial cells [53]. In summary, AMPs are tightly regulated during a host defense reaction. Much less is known about the inactivation or deactivation of AMPs. The antimicrobial activity of LL-37 in airway secretions is inhibited by bundle formation with F-actin and DNA [54]. Elastolytic cathepsins seem to be involved in degradation of hBD-2 and -3 [55]. Proteases secreted by common pathogenic bacteria degrade and inactivate the AMP LL-37 [56].

Functions of AMPs in the Respiratory Tract

Antimicrobial Activity

The antimicrobial activity of peptide antibiotics was deduced from in vitro tests assaying purified substances against microorganisms. AMPs have broad spectrum activity against gram-positive and gram-negative bacteria, as well as against fungi and enveloped viruses. Minimum inhibitory concentrations of the peptides range from 0.1 to 10 μg/ml. Human α-defensins may also be involved in defense against human immunodeficiency virus (HIV) [57]. The antimicrobial activity is based on several mechanisms that are described in more detail in other chapters of this book. In most cases, interactions between the peptide and surface membranes of the target organisms are considered to be responsible for activity. Individual antimicrobial components of the airway surface act synergistically against microorganisms.

Recently, several groups have published results that provide proof of the host defense function of AMPs in living organisms. Indirect in vivo evidence for the host defense function of AMPs came from a study in a murine model with a disrupted gene for matrilysin (metalloprotease 7). Mice with missing matrilysin were more susceptible to infection with enteropathogens [58]. Studies in a human bronchial xenograft model revealed decreased antimicrobial activity of ASF after inhibition of hBD-1 transcription by antisense oligonucleotides [9]. Mice deficient in the AMP β-defensin-1 (mBD-1) revealed delayed clearance of *Haemophilus influenzae* from lung [59] or higher bacterial load in the urinary tract [60]. Mice with deleted CRAMP, the murine homologue of LL-37, showed more prominent infection after cutaneous inoculation of bacteria [61]. In reverse, the overexpression of LL-37 by viral gene transfer resulted in augmentation of innate host defense in a bronchial xenograft model of cystic fibrosis and in murine models of pneumonia and septic shock [12, 62]. The transgenic expression of a human intestinal defensin in mice protected against enteric salmonellosis [63]. The existence of resistance mechanisms rendering bacteria less susceptible to the activity of AMPs is suggestive of a host defense function of these molecules in vivo. Increase of the phosphocholine content of cell walls of *H. influenzae* decreases bacterial susceptibility to LL-37 [64]. When exposed to the environment found in the airways of cystic fibrosis patients, *Pseudomonas aeruginosa* is able to modify the structure of the lipopolysaccharide (LPS) of the outer membrane [65]. The alterations of the endotoxin decrease the susceptibility of these bacteria to cationic AMPs.

Inflammation, Angiogenesis, and Cell Function

AMPs may have a variety of other biological effects beside their antimicrobial activity. Based on their membrane activity, AMPs have a concentration dependent toxicity towards eukaryotic cells. High concentrations of α-defensins have been described in secretions of patients with cystic fibrosis [66] and chronic bronchitis [67], where these substances likely contribute to the overwhelming inflammation. α-Defensins induce IL-8 production by lung epithelial cells [68]. The cellular damage by α-defensins is likely to be augmented by defensin-induced lysis of epithelial cells [69] or by binding of α-defensins to protease inhibitors of the serpin family such as alpha-1-antitrypsin [70]. When instilled into the respiratory tract of animals, they mediate an acute inflammatory response [71]. α-Defensins also boost the development of an acquired immune response through help provided by CD4+ Th1- and Th2-type helper cytokines, foster B and T cell interactions [72], and induce the secretion of secretory leukocyte proteinase inhibitor (SLPI) from epithelial cells [73].

Beside these nonspecific effects, AMPs bind to specific receptors and are involved in the regulation of chemotaxis, wound healing, angiogenesis, vascularization and epithelialization. These activities are summarized in table 1. Many AMPs are chemoattractants for leukocytes and other immune cells. HNP-1–3, hBD-1, hBD-2, hBD-3, mBD-2 and mBD-3 have been found to be chemotactic for immature dendritic cells [74]. Human α-defensins also chemoattract lymphocytes [75, 76]. Murine β-defensin 2 linked to a fusion protein has not only a chemotactic effect on CCR6 expressing immature dendritic cells [77], but acts directly on immature dendritic cells as an endogenous ligand for TLR-4 [47]. This interaction results in maturation of the dendritic cells including the upregulation of CD40, CD80, CD86, MHC class II and the chemokine receptor CCR7. hBD-1 and hBD-2 were found to bind to a chemokine receptor known as CCR6 [78]. This receptor is expressed on immature dendritic

Table 1. Antimicrobial peptides of the human respiratory tract and their binding receptors

Peptide family	Peptide name	Cellular source	Receptor	Function
α-Defensin	HNP-1-4	neutrophil	not identified	antibiotic; IL-8 secretion from epithelial cells; chemoattraction of T-cells, dendritic cells, monocytes; increased proliferation of epithelial cells; activation of mast cells
	HD5-6	airway epithelial cell	not identified	antibiotic
β-Defensin	hBD-1-4	airway epithelial cell, macrophage, type II pneumocyte	CCR6 (hBD-1, hBD-2); not identified (hBD-3, hBD-4)	antibiotic; chemoattractant for T cells, dendritic cells, and macrophages; activation of mast cells
Cathelicidin	LL-37/ hCAP-18	airway epithelial cell, macrophage, type II pneumocyte, neutrophil, lymphocyte	FPRL-1	antibiotic; chemoattractant for neutrophils, macrophages, and lymphocytes; activation of mast and epithelial cells angiogenic
Not identified	granulysin	cytolytic T lymphocytes and natural killer (NK) cells	not identified	antibiotic

and memory T cells (CD4+/CD45RO+) and consequently these findings are interpreted in favor of a link between innate and adaptive immune mechanisms mediated by defensins. hBD-3 and hBD-4 chemoattract monocytes by mechanisms that have not yet been clarified [17, 18]. HNP 1–3 are chemotactic for monocytes via a G protein-coupled receptor [75, 76]. hBD-2 directly and specifically induces mast cell migration through a pertussis toxin-sensitive and phospholipase C-dependent pathway [79]. LL-37 was found to bind to formyl peptide receptor like 1 (FPRL1), a promiscuous receptor expressed on a variety of cells including neutrophils, monocytes, and lymphocytes [80]. By activation of this G-protein coupled receptor, LL-37 attracts neutrophils, monocytes, and CD4 T cells and activates mast cells [81]. In human primary monocytes but not in B or T lymphocytes, LL-37 activates the extracellular signal-regulated kinase (ERK) and p38 kinase via a G protein-coupled receptor-independent mechanism not yet determined [82]. LL-37 also leads to IL-1 beta processing and release form monocytes [83]. This effect appears to be mediated by the P2X receptor. LL-37 also influences dendritic cell differentiation and dendritic cell induced T cell polarization [84]. Dendritic cells generated from blood monocytes under the influence of LL-37 have significantly up-regulated endocytic capacity, modified expression of phagocytic receptors, enhanced costimulatory molecule expression and secretion of Th-1 inducing cytokines [84]. The biological impact of these receptor mediated activities is completely unclear. Most AMPs are active at relatively high concentrations. It is unclear whether AMPs act as principal ligands for the mentioned receptors or whether they modulate the activities of other ligands. AMPs are also involved in wound healing and proliferation. Human neutrophil defensins induce the proliferation of airway epithelial cell lines via an EGF receptor-independent, MAP kinase signaling pathway [85]. Furthermore, neutrophil defensins enhance lung epithelial wound closure and mucin gene expression in vitro [86]. LL-37 activates human airway epithelial cells by activation of the mitogen-activated protein kinase (MAPK)/extracellular signal-regulated kinase (ERK) and increased release of IL-8 [87]. In vitro data suggest that LL-37 transactivates the EGF receptor involving metalloproteinase-mediated cleavage of membrane-anchored EGF receptor ligands [87]. LL-37 also plays a role in wound closure and re-epithelialization of human skin [88]. LL-37 stimulates angiogenesis in endothelial cells by activation of FPRL1 resulting in increased proliferation and formation of vessel-like structures in vitro [89]. Application of LL-37 results in neovascularization in the chorioallantoic membrane assay and in a rabbit model of hind-limb ischemia. Mice lacking CRAMP have decreased wound vascularization. PR-39 stimulates angiogenesis by binding to the alpha 7 subunit of the 26S proteasome and modulation of the ubiquitin-proteasome pathway without affecting overall proteasome activity [90]. Further, PR-39 is chemoattractive for neutrophils in a calcium dependent and pertussis toxin inhibitable reaction and contributes to wound healing by stimulating the expression of syndecans, cell surface heparane sulfate proteoglycans [91]. Interestingly, it has been found that the IFN-inducible ELR$^-$ CXC chemokine display antimicrobial activity [92]. Recently more than 17 human cytokines were identified to exhibit antimicrobial activity in vitro, including CCL20/MIP-3alpha [93], thus highlighting a structure-function relationship between AMPs and chemokines. Taken together, vertebrate AMPs have a variety of additional functions beside their microbicidal function. The impact of these non-microbicidal functions on the pathogenesis of diseases is completely speculative. The non-microbicidal functions offer interesting opportunities to investigate the roles of AMPs in inflammatory diseases, however, might also cause side effects when these peptides are used as therapeutics.

Role of AMPs in Pulmonary Disease

AMPs may have a role in a variety of infectious and inflammatory diseases of the lung. Based on their functions, several pathogenic models are relevant.

1. Inborn or acquired deficiencies of AMPs result in loss of function and subsequent increased susceptibility to infections. Only very limited data on states of decreased activity of AMPs are available. In morbus Kostmann [94], a severe congenital neutropenia, periodontal disease has been linked with the deficiency of AMPs in neutrophils [95]. A deficiency in the expression of AMPs may account for the susceptibility of patients with atopic dermatitis to skin infection with *Staphylococcus aureus* [51]. Chronic inflammation seems to be related with depressed expression of AMPs [52, 53, 96].
2. Several pulmonary diseases are associated with inflammation that results in many cases in overexpression of AMP genes. Based on the receptor-mediated functions of AMPs, increased concentrations may have a proinflammatory effect. Several examples are given below.
3. Polymorphisms of genes of AMPs may predispose for the development of pulmonary diseases. Several polymorphisms have been found in genes of β-defensins [97–100]. Polymorphisms of hBD-1 are associated with oral *Candida* carriage [101]. Variable numbers of copies of defensin genes contribute to the genetic complexity of these peptides [102].

Pneumonia and Tuberculosis

Infections of the respiratory tract are one of the most common disease groups. High numbers of hospital acquired pneumonias and increasing numbers of infections with multiply resistant bacteria are prominent problems. Several studies have found increased concentrations of defensins during infectious pulmonary diseases such as neonatal and adult pneumonia [34, 103]. Elevated levels of α-defensins were found in patients with empyema [104]. Tuberculosis is an infectious disease that in most cases involves the lung. The number of pan-resistant *Mycobacterium tuberculosis* strains is increasing. Levels of defensins are increased in plasma or bronchioalveolar lavage fluid in pulmonary tuberculosis [105] and infections with *Mycobacterium avium-intracellulare* [106].

Cystic Fibrosis and Diffuse Panbronchiolitis

Cystic fibrosis and diffuse panbronchiolitis are characterized by chronic infection associated with overwhelming neutrophilic inflammation. Cystic fibrosis (CF) is caused by a genetic defect of the CF transmembrane conductance regulator (CFTR). CF represents a model disease for defects of the innate host defense and research on this disease attracted significant attention to the field of AMPs in the late 1990s. Several hypotheses to explain the pathogenesis of CF lung disease have been proposed. The 'hypotonic airway surface fluid/antimicrobial substance' hypothesis proposes that the defects in CFTR result in elevated salt concentrations that inactivate antimicrobial substances. The core of this theory was the elevated NaCl content in CF ASF, an observation that is increasingly controversial. In contrast, the biogenesis or secretion of functional antimicrobial substances may be altered by intracellular defects in airway epithelial cells, as suggested by a salt-independent decrease of antimicrobial activity of CF airway secretions [107]. CF lung disease is clearly caused by a defect of the local innate immune system [108]. Whether AMPs have a direct role in the initial processes that link the defective CFTR with impaired host defense is unclear. In contrast, it appears that AMPs contribute to the overwhelming inflammatory activity. Several reports have found increased concentrations of β-defensins in CF airways [13, 107]. Also, α-defensins are found at increased levels [109]. Expression of β-defensins seems to be suppressed in the setting of chronic inflammatory CF lung disease [110, 111].

Diffuse panbronchiolitis is a chronic inflammatory lung disease of unknown origin that is phenotypically related to CF [112]. Neutrophil-derived defensins are elevated in the airways in diffuse panbronchiolitis and may be a marker of neutrophil activity in this disease [113]. Increased concentrations of β-defensins have been found in plasma and bronchoalveolar lavage fluid (BALF) of patients with this disease [114].

Asthma and Chronic Obstructive Pulmonary Disease

Asthma and chronic obstructive pulmonary disease (COPD) are obstructive pulmonary diseases that are characterized by airflow limitation and a chronic inflammatory process of the airways. These diseases have outstanding medical and economic importance. Inflammation in asthma is characterized by Th2 orchestrated inflammation, whereas in COPD, chronic smoke exposure (or other rare causes) results in neutrophil influx and activation of proteases. Based on their function as inflammatory mediators, AMPs likely are involved in the pathogenesis of these diseases. In contrast, no data about concentrations of AMPs in asthma and COPD are available. A polymorphism of the hBD-1 gene is found at higher frequency in COPD patients [98].

Adult Respiratory Distress Syndrome

Adult respiratory distress syndrome (ARDS) is a catastrophic inflammatory pulmonary disease that can be caused by a variety of conditions such as trauma, hypoxia, infection or intoxication. Neutrophil defensins are elevated in plasma and in BALF from patients with ARDS [115].

Pulmonary Fibrosis and Sarcoidosis

Pulmonary fibrosis is a descriptive term for a group of lung diseases characterized by various amounts of inflammation, destruction of lung parenchyma and replacement by fibrous materials accompanied by loss of pulmonary function. Plasma concentrations of α-defensins are raised in patients with pulmonary fibrosis [116]. Sarcoidosis is an inflammatory disease that mainly involves the pulmonary system. LL-37 is found to be upregulated in lungs of patients with this disease [26].

Conclusions

AMPs have emerged as multi-functional effector substances of the pulmonary innate immune system. On one side, convincing evidence has accumulated that these molecules are indeed endogenous antibiotics. On the other side, AMPs have additional functions as mediators of inflammation in the host defense scenario. A significant number of in vitro studies investigated these non-antimicrobial functions. Whether these activities are relevant in vivo is largely speculative. Therefore, at this time the role of AMPs in inflammatory lung disease is also speculative. Nevertheless, further studies on the receptor-mediated functions of AMPs are expected to provide many insights into human disease. Based on the presented data, AMPs qualify as potential innovative drugs. The broad spectrum of antimicrobial activity and the low incidence of bacterial resistance make them attractive candidates for prototypic novel antibiotics. If their role as inflammatory mediators is supported by in vivo evidence, appropriate strategies could be developed to influence human disease.

Acknowledgement

Studies in Dr. Bals' laboratory related to innate immunity of the respiratory tract are supported by grants of the Deutsche Forschungsgemeinschaft (Ba 1641/5–1, 6–1) and the Cystic Fibrosis Foundation (BALS03G0).

References

1 Smith J, Travis S, Greenberg E, Welsh M: Cystic fibrosis airway epithelia fail to kill bacteria because of abnormal airway surface fluid. Cell 1996;85:229–236.

2 Whitsett JA: Intrinsic and innate defenses in the lung: Intersection of pathways regulating lung morphogenesis, host defense, and repair. J Clin Invest 2002;109:565–569.

3 Bals R, Hiemstra PS: Innate immunity in the lung: How epithelial cells fight against respiratory pathogens. Eur Respir J 2004;23:327–333.

4 Bingle CD, Craven CJ: PLUNC: A novel family of candidate host defence proteins expressed in the upper airways and nasopharynx. Hum Mol Genet 2002;11:937–943.
5 Ganz T, Selsted ME, Szklarek D, Harwig SS, Daher K, Bainton DF, Lehrer RI: Defensins. Natural peptide antibiotics of human neutrophils. J Clin Invest 1985;76:1427–1435.
6 Selsted ME, Harwig SS, Ganz T, Schilling JW, Lehrer RI: Primary structures of three human neutrophil defensins. J Clin Invest 1985;76:1436–1439.
7 Bensch K, Raida M, Magert H-J, Schulz-Knappe P, Forssmann W-G: hBD-1: A novel b-defensin from human plasma. FEBS Lett 1995;368:331–335.
8 McCray PJ, Bentley L: Human airway epithelia express a beta-defensin. Am J Respir Cell Mol Biol 1997;16:343–349.
9 Goldman MJ, Anderson GM, Stolzenberg ED, Kari UP, Zasloff M, Wilson JM: Human beta-defensin-1 is a salt-sensitive antibiotic in lung that is inactivated in cystic fibrosis. Cell 1997;88:553–560.
10 Valore EV, Park CH, Quayle AJ, Wiles KR, McCray PB Jr, Ganz T: Human beta-defensin-1: An antimicrobial peptide of urogenital tissues. J Clin Invest 1998;101:1633–1642.
11 Harder J, Bartels J, Christophers E, Schroeder J-M: A peptide antibiotic from human skin. Nature 1997;387:861.
12 Bals R, Wang X, Wu Z, Freeman, Banfa V, Zasloff M, Wilson J: Human beta-defensin 2 is a salt-sensitive peptide antibiotic expressed in human lung. J Clin Invest 1998;102:874–880.
13 Singh P, Jia H, Wiles K, Hesselberth J, Liu L, Conway B, Greenberg E, Valore E, Welsh M, Ganz T, Tack B, McCray PJ: Production of beta-defensins by human airway epithelia. Proc Natl Acad Sci USA 1998;95:14961–14966.
14 Duits LA, Ravensbergen B, Rademaker M, Hiemstra PS, Nibbering PH: Expression of beta-defensin 1 and 2 mRNA by human monocytes, macrophages and dendritic cells. Immunology 2002;106:517–525.
15 Jia HP, Schutte BC, Schudy A, Linzmeier R, Guthmiller JM, Johnson GK, Tack BF, Mitros JP, Rosenthal A, Ganz T, McCray PB Jr: Discovery of new human beta-defensins using a genomics-based approach. Gene 2001;263:211–218.
16 Harder J, Bartels J, Christophers E, Schröder J-M: Isolation and characterization of human beta-defensin-3, a novel human inducible peptide antibiotic. J Biol Chem 2001;276:5707–5713.
17 Garcia JR, Jaumann F, Schulz S, Krause A, Rodriguez-Jimenez J, Forssmann U, Adermann K, Kluver E, Vogelmeier C, Becker D, Hedrich R, Forssmann WG, Bals R: Identification of a novel, multifunctional beta-defensin (human beta-defensin 3) with specific antimicrobial activity. Its interaction with plasma membranes of Xenopus oocytes and the induction of macrophage chemoattraction. Cell Tissue Res 2001;306:257–264.
18 Garcia JR, Krause A, Schulz S, Rodriguez-Jimenez FJ, Kluver E, Adermann K, Forssmann U, Frimpong-Boateng A, Bals R, Forssmann WG: Human beta-defensin 4: A novel inducible peptide with a specific salt-sensitive spectrum of antimicrobial activity. FASEB J 2001;15:1819–1821.
19 Schutte BC, Mitros JP, Bartlett JA, Walters JD, Jia HP, Welsh MJ, Casavant TL, McCray PB Jr: Discovery of five conserved beta-defensin gene clusters using a computational search strategy. Proc Natl Acad Sci USA 2002;99:2129–2133.
20 Tang Y-Q, Yaun J, Osapay G, Osapay C, Tran D, Miller C, Quellette A, Selsted M: A cyclic antimicrobial peptide produced in primate leukocytes by the ligation of two truncated alpha-defensins. Science 1999;286:498–502.
21 Zanetti M, Gennaro R, Romeo D: Cathelicidins: A novel protein family with a common proregion and a variable C-terminal antimicrobial domain. FEBS Lett 1995;374:1–5.
22 Gudmundsson GH, Agerberth B, Odeberg J, Bergman T, Olsson B, Salcedo R: The human gene FALL39 and processing of the cathelin precursor to the antibacterial peptide LL-37 in granulocytes. Eur J Biochem 1996;238:325–332.
23 Cowland J, Johnsen A, Borregaard N: hCAP-18, a cathelin/pro-bactenecin-like protein of human neutrophil specific granules. FEBS Lett 1995;368:173–176.
24 Bals R, Wang X, Zasloff M, Wilson JM: The peptide antibiotic LL-37/hCAP-18 is expressed in epithelia of the human lung where it has broad antimicrobial activity at the airway surface. Proc Natl Acad Sci USA 1998;95:9541–9546.

25 Agerberth B, Charo J, Werr J, Olsson B, Idali F, Lindbom L, Kiessling R, Jornvall H, Wigzell H, Gudmundsson GH: The human antimicrobial and chemotactic peptides LL-37 and alpha-defensins are expressed by specific lymphocyte and monocyte populations. Blood 2000;96: 3086–3093.
26 Agerberth B, Grunewald J, Castanos-Velez E, Olsson B, Jornvall H, Wigzell H, Eklund A, Gudmundsson GH: Antibacterial components in bronchoalveolar lavage fluid from healthy individuals and sarcoidosis patients. Am J Respir Crit Care Med 1999;160:283–290.
27 Sorensen OE, Follin P, Johnsen AH, Calafat J, Tjabringa GS, Hiemstra PS, Borregaard N: Human cathelicidin, hCAP-18, is processed to the antimicrobial peptide LL-37 by extracellular cleavage with proteinase 3. Blood 2001;97:3951–3959.
28 Sorensen OE, Gram L, Johnsen AH, Andersson E, Bangsboll S, Tjabringa GS, Hiemstra PS, Malm J, Egesten A, Borregaard N: Processing of seminal plasma hCAP-18 to ALL-38 by gastricsin: A novel mechanism of generating antimicrobial peptides in vagina. J Biol Chem 2003;20:20.
29 Zaiou M, Nizet V, Gallo RL: Antimicrobial and protease inhibitory functions of the human cathelicidin (hCAP18/LL-37) prosequence. J Invest Dermatol 2003;120:810–816.
30 Murakami M, Lopez-Garcia B, Braff M, Dorschner RA, Gallo RL: Postsecretory processing generates multiple cathelicidins for enhanced topical antimicrobial defense. J Immunol 2004;172: 3070–3077.
31 Stenger S, Hanson DA, Teitelbaum R, Dewan P, Niazi KR, Froelich CJ, Ganz T, Thoma-Uszynski S, Melian A, Bogdan C, Porcelli SA, Bloom BR, Krensky AM, Modlin RL: An antimicrobial activity of cytolytic T cells mediated by granulysin. Science 1998;282:121–125.
32 Oppenheim FG, Xu T, McMillian FM, Levitz SM, Diamond RD, Offner GD, Troxler RF: Histatins, a novel family of histidine-rich proteins in human parotid secretion. Isolation, characterization, primary structure, and fungistatic effects on *Candida albicans*. J Biol Chem 1988;263: 7472–7477.
33 Cole AM, Kim YH, Tahk S, Hong T, Weis P, Waring AJ, Ganz T: Calcitermin, a novel antimicrobial peptide isolated from human airway secretions. FEBS Lett 2001;504:5–10.
34 Schaller-Bals S, Schulze A, Bals R: Increased levels of antimicrobial peptides in tracheal aspirates of newborn infants during infection. Am J Respir Crit Care Med 2002;165:992–995.
35 Harder J, Meyer-Hoffert U, Teran LM, Schwichtenberg L, Bartels J, Maune S, Schröder J-M: Mucoid *Pseudomonas aeruginosa*, TNF-alpha, and IL-1beta, but not IL-6, induce human beta-defensin-2 in respiratory epithelia. Am J Respir Cell Mol Biol 2000;22:714–721.
36 Duits LA, Nibbering PH, van Strijen E, Vos JB, Mannesse-Lazeroms SP, van Sterkenburg MA, Hiemstra PS: Rhinovirus increases human beta-defensin-2 and -3 mRNA expression in cultured bronchial epithelial cells. FEMS Immunol Med Microbiol 2003;38:59–64.
37 Tsutsumi-Ishii Y, Nagaoka I: NF-kappa B-mediated transcriptional regulation of human beta-defensin-2 gene following lipopolysaccharide stimulation. J Leukoc Biol 2002;71:154–162.
38 Krisanaprakornkit S, Kimball JR, Dale BA: Regulation of human beta-defensin-2 in gingival epithelial cells: The involvement of mitogen-activated protein kinase pathways, but not the NF-kappaB transcription factor family. J Immunol 2002;168:316–324.
39 Moon SK, Lee HY, Li JD, Nagura M, Kang SH, Chun YM, Linthicum FH, Ganz T, Andalibi A, Lim DJ: Activation of a Src-dependent Raf-MEK1/2-ERK signaling pathway is required for IL-1alpha-induced upregulation of beta-defensin 2 in human middle ear epithelial cells. Biochim Biophys Acta 2002;1590:41–51.
40 Wang X, Zhang Z, Louboutin JP, Moser C, Weiner DJ, Wilson JM: Airway epithelia regulate expression of human beta-defensin 2 through Toll-like receptor 2. FASEB J 2003;17:1727–1729.
41 Lu Z, Kim KA, Suico MA, Shuto T, Li JD, Kai H: MEF up-regulates human beta-defensin 2 expression in epithelial cells. FEBS Lett 2004;561:117–121.
42 Tomita T, Nagase T, Ohga E, Yamaguchi Y, Yoshizumi M, Ouchi Y: Molecular mechanisms underlying human beta-defensin-2 gene expression in a human airway cell line (LC2/ad). Respirology 2002;7:305–310.
43 Hertz CJ, Wu Q, Porter EM, Zhang YJ, Weismuller KH, Godowski PJ, Ganz T, Randell SH, Modlin RL: Activation of Toll-like receptor 2 on human tracheobronchial epithelial cells induces the antimicrobial peptide human beta defensin-2. J Immunol 2003;171:6820–6826.

44 Birchler T, Seibl R, Buchner K, Loeliger S, Seger R, Hossle JP, Aguzzi A, Lauener RP: Human Toll-like receptor 2 mediates induction of the antimicrobial peptide human beta-defensin 2 in response to bacterial lipoprotein. Eur J Immunol 2001;31:3131–3137.

45 Becker MN, Diamond G, Verghese MW, Randell SH: CD14-dependent lipopolysaccharide-induced beta-defensin-2 expression in human tracheobronchial epithelium. J Biol Chem 2000; 275:29731–29736.

46 Tsutsumi-Ishii Y, Nagaoka I: Modulation of human beta-defensin-2 transcription in pulmonary epithelial cells by lipopolysaccharide-stimulated mononuclear phagocytes via proinflammatory cytokine production. J Immunol 2003;170:4226–4236.

47 Biragyn A, Ruffini PA, Leifer CA, Klyushnenkova E, Shakhov A, Chertov O, Shirakawa AK, Farber JM, Segal DM, Oppenheim JJ, Kwak LW: Toll-like receptor 4-dependent activation of dendritic cells by beta-defensin 2. Science 2002;298:1025–1029.

48 Griffin S, Taggart CC, Greene CM, O'Neill S, McElvaney NG: Neutrophil elastase up-regulates human beta-defensin-2 expression in human bronchial epithelial cells. FEBS Lett 2003;546:233–236.

49 Agerberth B, Gunne H, Odeberg J, Kogner P, Boman HG, Gudmundsson GH: FALL-39, a putative human peptide antibiotic, is cysteine-free and expressed in bane marrow and testis. Proc Natl Acad Sci USA 1995;92:195–199.

50 Hase K, Eckmann L, Leopard JD, Varki N, Kagnoff MF: Cell differentiation is a key determinant of cathelicidin LL-37/human cationic antimicrobial protein 18 expression by human colon epithelium. Infect Immun 2002;70:953–963.

51 Ong PY, Ohtake T, Brandt C, Strickland I, Boguniewicz M, Ganz T, Gallo RL, Leung DY: Endogenous antimicrobial peptides and skin infections in atopic dermatitis. N Engl J Med 2002;347:1151–1160.

52 Nomura I, Goleva E, Howell MD, Hamid QA, Ong PY, Hall CF, Darst MA, Gao B, Boguniewicz M, Travers JB, Leung DY: Cytokine milieu of atopic dermatitis, as compared to psoriasis, skin prevents induction of innate immune response genes. J Immunol 2003;171:3262–3269.

53 Islam D, Bandholtz L, Nilsson J, Wigzell H, Christensson B, Agerberth B, Gudmundsson G: Downregulation of bactericidal peptides in enteric infections: A novel immune escape mechanism with bacterial DNA as a potential regulator. Nat Med 2001;7:180–185.

54 Weiner DJ, Bucki R, Janmey PA: The antimicrobial activity of the cathelicidin LL37 is inhibited by F-actin bundles and restored by gelsolin. Am J Respir Cell Mol Biol 2003;28:738–745.

55 Taggart CC, Greene CM, Smith SG, Levine RL, McCray PB Jr, O'Neill S, McElvaney NG: Inactivation of human beta-defensins 2 and 3 by elastolytic cathepsins. J Immunol 2003;171:931–937.

56 Schmidtchen A, Frick IM, Andersson E, Tapper H, Bjorck L: Proteinases of common pathogenic bacteria degrade and inactivate the antibacterial peptide LL-37. Mol Microbiol 2002;46:157–168.

57 Zhang L, Yu W, He T, Yu J, Caffrey RE, Dalmasso EA, Fu S, Pham T, Mei J, Ho JJ, Zhang W, Lopez P, Ho DD: Contribution of human alpha-defensin 1, 2, and 3 to the anti-HIV-1 activity of CD8 antiviral factor. Science 2002;298:995–1000.

58 Wilson CL, Ouellette AJ, Satchell DP, Ayabe T, Lopez-Boado YS, Stratman JL, Hultgren SJ, Matrisian LM, Parks WC: Regulation of intestinal alpha-defensin activation by the metalloproteinase matrilysin in innate host defense. Science 1999;286:113–117.

59 Moser C, Weiner DJ, Lysenko E, Bals R, Weiser JN, Wilson JM: Beta-defensin 1 contributes to pulmonary innate immunity in mice. Infect Immun 2002;70:3068–3072.

60 Morrison G, Kilanowski F, Davidson D, Dorin J: Characterization of the mouse beta defensin 1, Defb1, mutant mouse model. Infect Immun 2002;70:3053–3060.

61 Nizet V, Ohtake T, Lauth X, Trowbridge J, Rudisill J, Dorschner RA, Pestonjamasp V, Piraino J, Huttner K, Gallo RL: Innate antimicrobial peptide protects the skin from invasive bacterial infection. Nature 2001;414:454–457.

62 Bals R, Weiner DJ, Meegalla RL, Wilson JM: Transfer of a cathelicidin peptide antibiotic gene restores bacterial killing in a cystic fibrosis xenograft model. J Clin Invest 1999;103:1113–1117.

63 Salzman NH, Ghosh D, Huttner KM, Paterson Y, Bevins CL: Protection against enteric salmonellosis in transgenic mice expressing a human intestinal defensin. Nature 2003;422:522–526.

64 Lysenko ES, Gould J, Bals R, Wilson JM, Weiser JN: Bacterial phosphorylcholine decreases susceptibility to the antimicrobial peptide LL-37/hCAP18 expressed in the upper respiratory tract. Infect Immun 2000;68:1664–1671.

65 Ernst RK, Yi EC, Guo L, Lim KB, Burns JL, Hackett M, Miller SI: Specific lipopolysaccharide found in cystic fibrosis airway *Pseudomonas aeruginosa*. Science 1999;286:1561–1565.
66 Soong L, Ganz T, Ellison A, Caughey G: Purification and characterization of defensins from cystic fibrosis sputum. Inflamm Res 1997;46:98–102.
67 Panyutich AV, Panyutich EA, Krapivin VA, Baturevich EA, Ganz T: Plasma defensin concentrations are elevated in patients with septicemia or bacterial meningitis. J Lab Clin Med 1993;122: 202–207.
68 van Wetering S, Mannesse-Lazeroms SP, van Sterkenburg MA, Daha MR, Dijkman JH, Hiemstra PS: Effect of defensins on interleukin-8 synthesis in airway epithelial cells. Am J Physiol 1997;272:888–896.
69 van Wetering S, Mannesse-Lazeroms SP, Dijkman JH, Hiemstra PS: Effect of neutrophil serine proteinases and defensins on lung epithelial cells: modulation of cytotoxicity and IL-8 production. J Leukoc Biol 1997;62:217–226.
70 Panyutich AV, Hiemstra PS, van Wetering S, Ganz T: Human neutrophil defensin and serpins form complexes and inactivate each other. Am J Respir Cell Mol Biol 1995;12:351–357.
71 Zhang H, Porro G, Orzech N, Mullen B, Liu M, Slutsky AS: Neutrophil defensins mediate acute inflammatory response and lung dysfunction in dose-related fashion. Am J Physiol Lung Cell Mol Physiol 2001;280:947–954.
72 Lillard JW Jr, Boyaka PN, Chertov O, Oppenheim JJ, McGhee JR: Mechanisms for induction of acquired host immunity by neutrophil peptide defensins. Proc Natl Acad Sci USA 1999;96: 651–656.
73 van Wetering S, van der Linden AC, van Sterkenburg MA, De Boer WI, Kuijpers AL, Schalkwijk J, Hiemstra PS: Regulation of SLPI and elafin release from bronchial epithelial cells by neutrophil defensins. Am J Physiol Lung Cell Mol Physiol 2000;278:51–58.
74 Yang D, Biragyn A, Hoover DM, Lubkowski J, Oppenheim JJ: Multiple roles of antimicrobial defensins, cathelicidins, and eosinophil-derived neurotoxin in host defense. Annu Rev Immunol 2004;22:181–215.
75 Chertov O, Michiel DF, Xu L, Wang JM, Tani K, Murphy WJ, Longo DL, Taub DD, Oppenheim JJ: Identification of defensin-1, defensin-2, and CAP37/azurocidin as T-cell chemoattractant proteins released from interleukin-8-stimulated neutrophils. J Biol Chem 1996;271:2935–2940.
76 Yang D, Chen Q, Chertov O, Oppenheim JJ: Human neutrophil defensins selectively chemoattract naive T and immature dendritic cells. J Leukoc Biol 2000;68:9–14.
77 Biragyn A, Surenhu M, Yang D, Ruffini PA, Haines BA, Klyushnenkova E, Oppenheim JJ, Kwak LW: Mediators of innate immunity that target immature, but not mature, dendritic cells induce antitumor immunity when genetically fused with nonimmunogenic tumor antigens. J Immunol 2001;167:6644–6653.
78 Yang D, Chertov O, Bykovskaia S, Chen Q, Buffo M, Shogan J, Anderson M, Schroder J, Wang J, Howard O, Oppenheim J: Beta-defensins: Linking innate and adaptive immunity through dendritic and T cell CCR6. Science 1999;286:525–528.
79 Niyonsaba F, Iwabuchi K, Matsuda H, Ogawa H, Nagaoka I: Epithelial cell-derived human beta-defensin-2 acts as a chemotaxin for mast cells through a pertussis toxin-sensitive and phospholipase C-dependent pathway. Int Immunol 2002;14:421–426.
80 Yang D, Chen Q, Schmidt AP, Anderson GM, Wang JM, Wooters J, Oppenheim JJ, Chertov O: LL-37, the neutrophil granule- and epithelial cell-derived cathelicidin, utilizes formyl peptide receptor-like 1 (FPRL1) as a receptor to chemoattract human peripheral blood neutrophils, monocytes, and T-cells. J Exp Med 2000;192:1069–1074.
81 Niyonsaba F, Someya A, Hirata M, Ogawa H, Nagaoka I: Evaluation of the effects of peptide antibiotics human beta-defensins-1/-2 and LL-37 on histamine release and prostaglandin D(2) production from mast cells. Eur J Immunol 2001;31:1066–1075.
82 Bowdish DM, Davidson DJ, Speert DP, Hancock RE: The human cationic peptide LL-37 induces activation of the extracellular signal-regulated kinase and p38 kinase pathways in primary human monocytes. J Immunol 2004;172:3758–3765.
83 Elssner A, Duncan M, Gavrilin M, Wewers MD: A novel P2X7 receptor activator, the human cathelicidin-derived peptide LL37, induces IL-1 beta processing and release. J Immunol 2004;172:4987–4994.

84 Davidson DJ, Currie AJ, Reid GS, Bowdish DM, MacDonald KL, Ma RC, Hancock RE, Speert DP: The cationic antimicrobial peptide LL-37 modulates dendritic cell differentiation and dendritic cell-induced T cell polarization 10. J Immunol 2004;172:1146–1156.
85 Aarbiou J, Ertmann M, van Wetering S, van Noort P, Rook D, Rabe KF, Litvinov SV, Van Krieken JH, De Boer WI, Hiemstra PS: Human neutrophil defensins induce lung epithelial cell proliferation in vitro. J Leukoc Biol 2002;72:167–174.
86 Aarbiou J, Verhoosel RM, van Wetering S, De Boer WI, Van Krieken JH, Litvinov SV, Rabe KF, Hiemstra PS: Neutrophil defensins enhance lung epithelial wound closure and mucin gene expression in vitro. Am J Respir Cell Mol Biol 2003;18:18.
87 Tjabringa GS, Aarbiou J, Ninaber DK, Drijfhout JW, Sorensen OE, Borregaard N, Rabe KF, Hiemstra PS: The antimicrobial peptide LL-37 activates innate immunity at the airway epithelial surface by transactivation of the epidermal growth factor receptor. J Immunol 2003:171.
88 Heilborn JD, Nilsson MF, Kratz G, Weber G, Sorensen O, Borregaard N, Stahle-Backdahl M: The cathelicidin anti-microbial peptide LL-37 is involved in re-epithelialization of human skin wounds and is lacking in chronic ulcer epithelium. J Invest Dermatol 2003;120:379–389.
89 Koczulla R, von Degenfeld G, Kupatt C, Krotz F, Zahler S, Gloe T, Issbrucker K, Unterberger P, Zaiou M, Lebherz C, Karl A, Raake P, Pfosser A, Boekstegers P, Welsch U, Hiemstra PS, Vogelmeier C, Gallo RL, Clauss M, Bals R: An angiogenic role for the human peptide antibiotic LL-37/hCAP-18. J Clin Invest 2003;111:1665–1672.
90 Li J, Post M, Volk R, Gao Y, Li M, Metais C, Sato K, Tsai J, Aird W, Rosenberg RD, Hampton TG, Sellke F, Carmeliet P, Simons M: PR39, a peptide regulator of angiogenesis. Nat Med 2000;6: 49–55.
91 Gallo R, Ono M, Povsic T, Page C, Eriksson E, Klagsburn M, Burnfield M: Syndecans, cell surface heparan sulfate proteoglycans, are induces by a proline-rich antimicrobial peptide from wounds. Proc Natl Acad Sci USA 1994;91:11035–11039.
92 Cole AM, Ganz T, Liese AM, Burdick MD, Liu L, Strieter RM: Cutting edge: IFN-inducible ELR-CXC chemokines display defensin-like antimicrobial activity. J Immunol 2001;167:623–627.
93 Yang D, Chen Q, Hoover DM, Staley P, Tucker KD, Lubkowski J, Oppenheim JJ: Many chemokines including CCL20/MIP-3alpha display antimicrobial activity. J Leukoc Biol 2003;74:448–455.
94 Zeidler C, Welte K: Kostmann syndrome and severe congenital neutropenia. Semin Hematol 2002;39:82–88.
95 Putsep K, Carlsson G, Boman HG, Andersson M: Deficiency of antibacterial peptides in patients with morbus Kostmann: An observation study. Lancet 2002;360:1144–1149.
96 Fellermann K, Wehkamp J, Herrlinger KR, Stange EF: Crohn's disease: A defensin deficiency syndrome? Eur J Gastroenterol Hepatol 2003;15:627–634.
97 Circo R, Skerlavaj B, Gennaro R, Amoroso A, Zanetti M: Structural and functional characterization of hBD-1(Ser35), a peptide deduced from a DEFB1 polymorphism. Biochem Biophys Res Commun 2002;293:586–592.
98 Matsushita I, Hasegawa K, Nakata K, Yasuda K, Tokunaga K, Keicho N: Genetic variants of human beta-defensin-1 and chronic obstructive pulmonary disease. Biochem Biophys Res Commun 2002;291:17–22.
99 Vatta S, Boniotto M, Bevilacqua E, Belgrano A, Pirulli D, Crovella S, Amoroso A: Human beta defensin 1 gene: Six new variants. Hum Mutat 2000;15:582–583.
100 Dork T, Stuhrmann M: Polymorphisms of the human beta-defensin-1 gene. Mol Cell Probes 1998;12:171–173.
101 Jurevic RJ, Bai M, Chadwick RB, White TC, Dale BA: Single-nucleotide polymorphisms (SNPs) in human beta-defensin 1: High-throughput SNP assays and association with Candida carriage in type I diabetics and nondiabetic controls. J Clin Microbiol 2003;41:90–96.
102 Hollox EJ, Armour JA, Barber JC: Extensive normal copy number variation of a beta-defensin antimicrobial-gene cluster. Am J Hum Genet 2003;73:591–600.
103 Hiratsuka T, Nakazato M, Date Y, Ashitani J, Minematsu T, Chino N, Matsukura S: Identification of human beta-defensin-2 in respiratory tract and plasma and its increase in bacterial pneumonia. Biochem Biophys Res Commun 1998;249:943–947.
104 Ashitani J, Mukae H, Nakazato M, Taniguchi H, Ogawa K, Kohno S, Matsukura S: Elevated pleural fluid levels of defensins in patients with empyema. Chest 1998;113:788–794.

105 Ashitani J, Mukae H, Hiratsuka T, Nakazato M, Kumamoto K, Matsukura S: Elevated levels of alpha-defensins in plasma and BAL fluid of patients with active pulmonary tuberculosis. Chest 2002;121:519–526.

106 Ashitani J, Mukae H, Hiratsuka T, Nakazato M, Kumamoto K, Matsukura S: Plasma and BAL fluid concentrations of antimicrobial peptides in patients with Mycobacterium avium-intracellulare infection. Chest 2001;119:1131–1137.

107 Bals R, Weiner DJ, Meegalla RL, Accurso F, Wilson JM: Salt-independent abnormality of antimicrobial activity in cystic fibrosis airway surface fluid. Am J Respir Cell Mol Biol 2001;25:21–25.

108 Bals R, Weiner DJ, Wilson JM: The innate immune system in cystic fibrosis lung disease. J Clin Invest 1999;103:303–307.

109 Soong LB, Ganz T, Ellison A, Caughey GH: Purification and characterization of defensins from cystic fibrosis sputum. Inflamm Res 1997;46:98–102.

110 Dauletbaev N, Gropp R, Frye M, Loitsch S, Wagner TO, Bargon J: Expression of human beta defensin (HBD-1 and HBD-2) mRNA in nasal epithelia of adult cystic fibrosis patients, healthy individuals, and individuals with acute cold. Respiration 2002;69:46–51.

111 Chen CIU, Schaller-Bals S, Paul KP, Wahn U, Bals R: Beta-defensin and LL-37 in bronchalveolar lavage fluid of patients with cystic fibrosis. J Cystic Fibrosis 2004;3:45–50.

112 Yanagihara K, Kadoto J, Kohno S: Diffuse panbronchiolitis: Pathophysiology and treatment mechanisms. Int J Antimicrob Agents 2001;18:83–87.

113 Ashitani J, Mukae H, Nakazato M, Ihi T, Mashimoto H, Kadota J, Kohno S, Matsukura S: Elevated concentrations of defensins in bronchoalveolar lavage fluid in diffuse panbronchiolitis. Eur Respir J 1998;11:104–111.

114 Hiratsuka T, Mukae H, Iiboshi H, Ashitani J, Nabeshima K, Minematsu T, Chino N, Ihi T, Kohno S, Nakazato M: Increased concentrations of human beta-defensins in plasma and bronchoalveolar lavage fluid of patients with diffuse panbronchiolitis. Thorax 2003;58:425–430.

115 Ashitani J, Mukae H, Ihiboshi H, Taniguchi H, Mashimoto H, Nakazato M, Matsukura S: Defensin in plasma and in bronchoalveolar lavage fluid from patients with acute respiratory distress syndrome. Nippon Kyobu Shikkan Gakkai Zasshi – Jpn J Thorac Dis 1996;34:1349–1353.

116 Mukae H, Iiboshi H, Nakazato M, Hiratsuka T, Tokojima M, Abe K, Ashitani J, Kadota J, Matsukura S, Kohno S: Raised plasma concentrations of alpha-defensins in patients with idiopathic pulmonary fibrosis. Thorax 2002;57:623–628.

Robert Bals, MD, PhD
Department of Internal Medicine, Division of Pulmonology
Hospital of the University of Marburg, Baldingerstrasse 1
DE–35043 Marburg (Germany)
Tel. +49 6421 2866451, Fax +49 6421 2868987, E-Mail bals@mailer.uni-marburg.de

Kabelitz D, Schröder J-M (eds): Mechanisms of Epithelial Defense.
Chem Immunol Allergy. Basel, Karger, 2005, vol 86, pp 72–98

Bacterial Evasion of Innate Defense at Epithelial Linings

Mathias W. Hornef[a,b], *Staffan Normark*[b], *Birgitta Henriques-Normark*[b], *Mikael Rhen*[b]

[a]Institute for Medical Microbiology and Hygiene, University of Freiburg, Germany; [b]Microbiology and Tumor Biology Center (MTC), Karolinska Institutet, Stockholm, Sweden

Abstract

The evolution of multicellular organisms has been, and continues to be, paralleled by the evolution of the surrounding microbial flora. This intimate coexistence between higher host organisms and microbes has generated a myriad of adaptation strategies at both sides to cope with, or even benefit from the given conditions. On the host side, the development of an effective immune defence system allowed the maintenance of an astonishingly stable homeostasis on many body sites, and even the establishment of sterile surfaces at vulnerable anatomical sites. On the other hand, microbial diversity has led to the establishment of a large number of microbial life styles that allow persistence and proliferation in the presence of host defense mechanisms. The following review describes bacterial strategies to circumvent or modify host defenses that operate at the epithelial lining. It illustrates the enormous diversity of mechanisms that are part of the complex interplay between microbial organisms and the host. It also reflects the dramatic progress made in the understanding of the mammalian immune defense system which many times has been initiated by the surprising results from the study of microbial pathogenesis.

How to Circumvent Physical Removal from Body Surfaces

Most body surfaces are densely colonized with a variety of microorganisms. In fact, the total number of colonizing microbial cells by far outnumbers the cells that form the human body. This so-called 'normal flora' is well tolerated under healthy conditions and certainly is beneficial to the host organism [1]. The skin, upper respiratory tract, lower urogenital tract and parts of the intestinal tract have evolved strategies to sustain permanent colonization. In contrast, other anatomical sites are maintained as sterile surfaces as a prerequisite for the essential physiological organ function such as air exchange in the lung, fluid and substrate excretion in the kidney, and reproduction in the upper female genital tract. Still, also heavily colonized mucosal surfaces are subjected to continuous physical forces and immune surveillance mechanisms removing living and inorganic particles. Mechanical clearance is mainly achieved through continuous washing with fluid to capture and remove possible contaminating microorganisms (such as seen in the urinary tract), through peristalsis (such as in the gastrointestinal tract), or through directed mucociliary transport (such as seen in the respiratory tract). It follows that microbial colonization and multiplication require firm attachment to host cell surfaces. Not surprisingly, the ability to express adhesive organelles or adhesions belonged to the first bacterial virulence factors identified [2].

The classical bacterial adhesin consists of elongated proteinaceous filaments protruding out from the bacterial surface, sometimes at lengths of micrometers [3, 4]. Such adhesins are referred to as fimbriae or pili built up of protein subunits called fimbrillins or pilins. While these organelles might be rather hydrophobic in character and possibly able to mediate more or less unspecific hydrophobic interactions, the adhesins usually function through an astonishingly specific binding to selected host cell surface structures [5]. In most instances this specificity is created by the adhesion molecule that is placed on the tip of the long structural extensions to allow binding from a distant location [6]. The recognized structures mainly represent characteristic surface carbohydrate residues such as blood-group antigens, and binding anchors the bacteria to selected surfaces [7]. Consequently, the receptor-recognition repertoire determines host specificity but also tissue tropism. Recognition of polymeric structures and the simultaneous expression of multiple adhesions facilitate strong, polyvalent binding that resists physical removal. Binding to carbohydrate structures impairs the host's possibilities to mutate the recognized structure and avoid microbial attachment and colonization.

One illustrating example of pili-mediated colonization comes from uropathogenic *Escherichia coli* (UPECs) causing urinary tract infections, a frequent clinical condition with significant morbidity particularly in women. Such strains are usually expressing a special class of pili termed either Pap pili or P-fimbriae that recognize the Galβ1–4Gal disaccharide present on glycosphingolipids and included in the P-blood-group antigen system [8]. Applying primate-based infection models, it has been shown that bacteria lacking Pap pilus expression become attenuated illustrating the importance of adhesive structures for microbial pathogenesis [9, 10].

Newer studies have revealed an additional important aspect of pilus-mediated binding: the initial distant attachment is followed in some cases by contraction of the long pili structures that brings the microbe closer to the epithelial lining and allows more intimate contact for example via non-piliated adhesive structures [11]. In fact, some bacteria such as for example uropathogenic *E. coli* (UPECs) or the enteropathogenic *Yersinia* (*Y. enterocolitica* and *Y. peudotuberculosis*) as well as *Listeria monocytogenes* employ piliated or non-piliated adhesions for cellular invasion (discussed later in this chapter) [12–14].

Perhaps the most striking example of a bacteria-driven adhesion process is provided by an enteropathogenic variant of *E. coli* (EPEC). This important causative agent of diarrhea provides its own receptor, Tir, which is translocated into the epithelial cell membrane to subsequently facilitate bacterial binding [15]. The application of the receptor requires a special so-called type III protein secretion system [16, 17]. The system resembles in complexity the flagellar machinery of gram-negative bacteria, and constitutes literally a syringe structure by which bacterial virulence and effector proteins can be injected into the host cell cytoplasm across the mammalian membrane. Type III secretion systems are by far not restricted to EPEC, but as we shall see rather represent a conserved bacterial strategy for steering the infection pathogenesis.

Receptors recognized by bacterial adhesins are not by definition cell surface molecules. A parallel and evidently broadly used strategy is high affinity binding to extracellular matrix (ECM) proteins, typically fibronectin, laminins and collagens [18]. Due to the interaction of the ECM with cellular structures, such binding indirectly also mediates adhesion to host cell surfaces. For example, *Staphylococcus aureus*, an important cause of wound infection and abscess formation is able to bind fibronectin [19], the extracellular ligand of β1 integrins, which in turn are ubiquitously expressed on epithelial cells. Binding to fibronectin and other extracellular matrix proteins is mediated by cell wall-anchored proteins so called MSCRAMMs (microbial surface components recognizing adhesive matrix molecules) using repeats in the C-terminal region of the bacterial protein [20]. Binding to matrix protein through fimbrial organelles has also been described for the enteric bacteria *Salmonella* and *E. coli* [21]. Similar to the binding mediated by cell-surface structures, cell adhesion by soluble factors such as fibronectin-binding proteins (FnBPs) may also play a role in bacterial internalization into non-phagocytic cells (discussed below).

Bacteria such as *Pseudomonas aeruginosa* or *Staphylococcus epidermidis* are able to produce an extracellular matrix that forms a tight substrate called biofilm embedding the incorporated bacteria [22, 23]. This specialized matrix firmly attaches the microorganisms to the surface and protects them from the action of soluble antimicrobial factors. The exopolysaccharide intercellular adhesion (PIA) expressed on the surface of *S. epidermidis* anchors the extracellular matrix. Mutants lacking PIA expression are more susceptible to killing by polymorphonuclear leukocytes and the action of antimicrobial peptides as compared to the wild type [24]. Biofilm formation is particularly observed during foreign body infections such as indwelling catheters or plastic implants and treatment in most cases requires removal of the foreign material due to the insufficient penetration of antibacterial drugs into the biofilm. Alternatively, antimicrobial resistance may be specifically upregulated in bacteria grown in biofilm, but not during free-living (planctonic) growth [25, 26].

Clearly, firm attachment on epithelial cells resists physical clearance of the adhering bacterium. Another more aggressive strategy to overcome host clearance mechanisms is invasion into the intracellular space to persist at this remote localization. For example, uropathogenic *E. coli* (UPECs) have been demonstrated to trigger internalization of the microbe in epithelial cells by adhesion through the mannose binding lectin on type 1 pili. Once internalized the microbe organizes itself in large multicellular structures called 'pods', which appear to represent an intracellular mode of biofilm formation [27, 28]. Persistence at this remote localization might explain the recurrent character of urinary infections despite adequate antibiotic treatment. Nevertheless, the host has evolved mechanisms to clear this type of infection. Apoptosis of the most upper epithelial cell layer and exfoliation together with the attached microorganisms have been shown to represent an important mechanism to clear bacterial colonization in the lower urinary tract [29].

A similar host innate defence strategy might represent high cell turnover, short life span, and continuous shedding of the epithelial cells along the gastrointestinal tract. Continuous renewal of the most upper epithelial layer would effectively diminish close attachment and persistent colonization of the surface barrier by the dense enteric microflora. It might thereby help to maintain the essential function of the epithelium, transport of nutrients and liquid across the epithelial lining, allowing only a moderate bacterial colonization. Similarly, apoptosis of bronchial epithelial cells was shown to be essential for host survival in a pulmonary *P. aeruginosa* infection model. *P. aeruginosa* is a common cause of pneumonia in patients treated in intensive care units as well as patients with the inherited disease cystic fibrosis. Apoptosis and shedding of infected cells might prevent bacterial colonization and systemic dissemination [30]. Pathogen-induced inhibition of cell differentiation, proliferation, and apoptosis might on the other hand represent strategies to counteract continuous epithelial renewal and allow microbial colonization and proliferation.

Another strategy to circumvent removal by mucosal clearance mechanisms is seen during infection with *Bordetella pertussis*. This important upper respiratory pathogen, the causative agent of whooping cough, secretes bacterial cell wall constituents (muropeptides) during microbial attachment. Secreted muropeptide fragments in turn induce the production of nitric oxide (NO), most likely by activating the intracellular NOD/Card family of receptors. NO secretion mediates ciliostasis and allows persistent bacterial colonization, explaining the chronic character of *B. pertussis* infection [31].

Overcome Space and Nutrient Deprivation

While attachment certainly aids anchoring to host surfaces, any successful colonization would still depend on the presence of space and available nutrients. Both items are at limited access by the presence of a dense and highly dynamic normal flora on many surfaces. Metabolic fitness and microbe-microbe interactions are therefore intrinsic features of microbial pathogenesis and the physiological microbial flora therefore provides significant protection. This is best illustrated by the opportunistic pathogen *Clostridium difficile*, an important causative agent of hospital-acquired diarrhea and pseudomembranous colitis [32]. Infection with this ubiquitous spore forming bacterium is seen in patients with alterations of the normal flora, most frequently generated by treatment with broad-spectrum antibiotics but also in the context of more subtle changes in the intestinal ecology. Indeed, the importance of even minor changes in the gastrointestinal luminal milieu is illustrated by the finding that dietary factors such as oligofructose, inulin, or calcium consumption in murine models significantly influence the susceptibility of the host towards infection with enteropathogenic bacteria such as *Listeria monocytogenes* and *Salmonella enterica* serovar Enteritidis (*S. enteritidis*). The enhanced susceptibility might result from the composition of the microbial flora [33, 34].

Many microbial organisms are known to produce and secrete antibacterial substances, so called bacteriocins, presumably with the purpose to suppress the surrounding microflora and to gain access to epithelial surface [35]. Since commensal as well as pathogenic bacteria display different levels of susceptibility, the secretion of different bacteriocins at a given location may critically determine the local microflora composition. In fact, many modern antibiotic drugs structurally rely on substances that have originally been isolated from fungi or bacteria.

Although limited space and nutrient availability warrants microbial competition among members of the mucosal flora, also cooperation between different bacterial genera as been observed. For example, multigeneric polymicrobial plaques have been found in the periodontal pocket of patients with gingivitis and periodontitis. The anaerobe bacterium *Fusobacterium nucleatum* has been frequently isolated in both clinical conditions and seems to act as a bridge to facilitate attachment of different bacteria to the site of infection. These 'coaggregations' might provide an ecological advantage for the attachment to epithelial surfaces and thus present an example of bacterial synergy in the pathogenesis of periodontal disease [36, 37].

While it may be difficult to pinpoint the specific nutrient limitations that are growth restricting for individual colonizers or pathogens, the availability of free iron is certainly one factor. To avoid microbial access to host iron, it is transported and stored in the host 'chelated' in specialized carrier proteins with high ligand affinity such as ferritin or transferrin. In addition, lactoferrin is produced at the site of infection to additionally reduce the amount of free iron. However, most pathogenic bacteria release own iron trapping molecules, so called siderophores that show even higher binding affinity to iron in competition with host iron carriers [38]. Uptake of iron-loaded siderophores thus provides the main external microbial source of iron during infection. In addition, some pathogenic bacteria such as *Neisseria* spp. and *Hemophilus* spp. gain access to iron by binding host transferrin to receptors on the bacterial outer membrane, so called transferring-binding proteins (Tbp) [39]. The importance of the competition for iron is illustrated by the recent discovery that the mammalian host has evolved a carrier molecule called neutrophil gelatinase-associated lipocalin (NGAL) that in turn binds microbial siderophores to recover the complexed iron [40]. In addition, the free serum iron concentration is significantly reduced during infection by interleukin (IL)-6-induced secretion of a soluble factor from the human liver called hepcidin that inhibits cellular iron excretion [41]. The clinical relevance of low iron availability during infection is demonstrated by the deleterious effect of iron substitution in patients suffering from infection with *Mycobacterium tuberculosis* [42].

Thus, a picture emerges in which microbial colonization reflects a complex process that is influenced by metabolic, competitive, and cooperative properties in addition to what has been described as virulence factors (e.g. specific adhesions). The importance of epithelial colonization for many microbial pathogens is illustrated by the high prevalence of bacteria such as *Haemophilus influenzae*, *Streptococcus pneumoniae*, *Streptococcus pyogenes*, or *Neisseria meningitidis* among the microbial flora in the upper respiratory tract of healthy individuals.

Resisting the Low-pH Defense Barrier

In addition to ecological competition, colonization of some anatomical body sites is inhibited by the creation of a hostile, growth-limiting environment. One of the most fundamental environmental factors in this respect is the creation of an acidic pH, such as seen at the epithelial surface of the stomach and lower female genital tract, and the phagolysosomes of phagocytic cells.

Interestingly, acid production on the epithelial surface of the vagina of fertile women appears to be provided by carbohydrate fermentation by the colonizing apathogenic bacterium *Lactobacillus* [43]. This in turn prevents colonization by other potentially harmful pH sensitive microbial organisms [44], once again high-lightening the role of the normal microflora in the host innate defence. Not surprisingly commonly isolated pathogens in bacteria vaginosis, *Gardnerella vaginalis* and *Peptostreptococcus anaerobius*, are relatively acid resistant which allows them to gain access to this anatomical site.

In contrast, the low pH in the gastric lumen is produced by energy consuming ion transport across the epithelial lining. This creates a strongly acidic environment in the gastric lumen with pH values of 1–2 and contributes to a remarkably low microbial colonization within the upper gastrointestinal tract. It also represents a significant protection from gastric and intestinal microbial overgrowth and the transmission of oral infection. Hypochlorhydria is in fact known to predispose to non-typhoidal salmonella infections [45, 46] and cholera [45, 47]. Interestingly, many bacteria are not uniformly susceptible to pH but instead express a tolerance level that depends on the previous cultivation conditions [48]. Briefly, the bacteria may show a rather significant acid sensitivity if grown in rich medium at neutral pH and rapidly exposed to a drop in pH. If the culture in contrast is exposed to mild acidity in proper medium, it may subsequently tolerate much steeper drops in pH. For example, growth of the enteric bacterium *Salmonella enterica* serovar Typhimurium (*S. typhimurium*) in moderately low pH minimal medium induces the expression of sets of proteins, many of which evidently adapt the bacteria to harsher conditions. Interestingly, many of the gene regulatory factors, such as Fur, PhoP, RpoS and H-NS, involved in the induction of an acid tolerance are also needed for virulence expression [48, 49], and selected genes induced by the response seem to play a role in Salmonella virulence [50]. Thus, induction of an acid tolerance response, either in the environment or in the gastric juice, might actually provoke virulence [51].

Nevertheless, one prominent bacterial organism, *Helicobacter pylori*, is able to colonize and persist within the human gastric mucosa. *H. pylori* causes chronic gastric inflammation, a condition that may eventually lead to ulcer disease and the development of gastric cancer. Prior to colonization of the mucosal surface (which has a less acidic pH), it must pass the extremely acidic environment of the gastric lumen. Survival in this hostile milieu seems to be facilitated by the synthesis of large amounts of urease that maintains the pH homeostasis through the production of ammonia from exogeneous urea [52]. *H. pylori* defective in the production of ammonia exhibits decreased survival and colonization within the gastric lumen in a gerbil infection model [53]. Also, the exoprotein VacA of *H. pylori* has been described to promote urea diffusion across the apical epithelial membrane and thereby enhance the supply with urea, critical for the buffering capacity generated by urease activity [54]. Interestingly, the same exoprotein was also shown to enhance alkaline secretion from gastric epithelial cells [55] as if to prepare the mucosal milieu for proper microbial growth.

Finally, escape from the hostile environment and move towards nutrition is also enhanced by microbial motility. Only motile, flagellated *H. pylori* strains were able to colonize in an animal model [56], reflecting the need to reach protected sites within the gastric lumen. Similarly, an early observation noted that flagellin represented an essential virulence factor for *Vibrio cholerae*, the causative agent of cholera, an epidemic enteritis with major socioecological importance worldwide [57].

Avoid Protease Mediated Destruction and Opsonization

The host possesses several means of coping with intruders that might have overcome initial attempts of removal and physicochemical barriers. Specific host effector molecules target approaching microbial organisms to directly eliminate them or activate host immune effector cells. Enzymes such as lysozyme are produced at high concentrations in lung and intestinal tissue and provide significant protection from bacterial infection. Interestingly, a recent report described inhibition of lysozyme activity by a surface protein of *Streptococcus pyogenes*, an important cause of upper respiratory tract and soft-tissue infections [58].

In addition, immunoglobulins (mainly of the IgA type) are translocated across the epithelium to protect mucosal linings. Binding of immunoglobulins to the microbial surface immobilizes the microbial organism and facilitates cellular uptake by professional immune cells via specific immunoglobulin receptors (Fc receptors). It further promotes deposition of complement factors on the microbial surface (discussed below). However, highy-affinity binding of immunoglobulins requires previous exposure and activation of the adaptive immune system. Microbial pathogens may benefit from the time required to mount a specific acquired response, either by undergoing antigenic variation, or by escaping into protected sites before a significant humoral response becomes mounted [59]. Nevertheless also low affinity binding by natural polyreactive IgA antibodies present on mucosal surfaces might significantly contribute to the epithelial defense [60].

Some bacteria are able to proteolytically degrade secretory immunoglobulins. This strategy is particularly seen in bacteria that colonize the upper respiratory tract such as for example *Haemophilus influenzae* or *S. pneumoniae*, both causative agents of respiratory tract infections such as otitis, sinusitis and, if capsulated, also of meningitis [61]. *Streptococcus pyogenes*, another causative agent of upper respiratory infections (such as acute pharyngitis) but also skin and soft-tissue infections, tightly binds the Fc-portion of the immunoglobulin molecule via its fibronection-binding protein 1 (SfbI) and thereby inhibits Fc receptor mediated internalization and complement activation [62]. Similarly, *Staphylococcus aureus*, responsible for hospital acquired pneumonia and sepsis, but also for a series of skin derived infections, tightly binds the Fc-portion of the immunoglobulin via expression of protein A and might thereby inhibit Fc receptor mediated internalization [63].

Activation of the complement system, an evolutionary old defense machinery occurs following antibody binding (classical pathway), through an antibody-independent binding of the mannan-binding lectin (MBL pathway) or via deposition of the early complement factors on the microbial surface (alternative pathway). Stimulation of a cascade of proteolytically activated effector molecules subsequently results in the generation of proinflammatory mediators (particularly the anaphylatoxins C3a and C5a) as well as the formation of a large molecular complex, the membrane attack complex (MAC), which is able to perforate microbial membranes. In addition, binding of complement factors on a microbial surface (opsonization) strongly enhances the internalization by professional immune cells via complement receptors.

Microbial organisms have evolved strategies at multiple levels to avoid recognition or circumvent destruction by the complement system. A variety of soluble and membrane bound host regulatory factors prevent the deposition of complement on host cell surfaces. Shielding from complement activity is obtained by the production of a hyaluronic acid capsule such as seen in *Streptococcus pneumoniae* or capsule-like structures such as the extracellular protein YadA from *Yersinia* [64]. Another general theme is to bind complement regulatory factors. For example, *Streptococcus agalactiae* (also called group B streptococcus) an important pathogen of newborn sepsis and maternal puerperal fever, as well as *S. pneumoniae* both are able to bind the regulatory factor H via specific bacterial surface proteins. Binding confers protection from complement deposition [65]. *Streptococcus pyogenes*, a common agent of skin, pharyngeal, and soft-tissue infections, employs the negative complement regulator C4b binding protein (C4bp) from the host by expression of a bacterial C4b homologue, the so-called M protein [66]. In addition, a surface protein of *S. pyogenes*, streptococcus inhibitor of complement (SIC), prevents the formation of the membrane attack complex by binding of the C5b-C7 complex [67]. *S. pyogenes* is also able to inactivate the potent chemoattractive complement factor C5a using a surface bound peptidase [68]. Thus, some bacteria may express a variety of mechanisms simultaneously to counteract a specific host defense strategy such as complement activation, clear evidence for the clinical importance of this defensive arm in host immunity.

In this context, another surprising strategy has been described for the respiratory pathogen *Haemophilus influenzae. H. influenzae* expresses a sialyltransferase that transfers sialic acid from host cells to terminal sugar residues of the bacterial outer membrane constituent lipopolysaccharide (LPS) [69, 70]. Sialylation mimics host structures and inhibits complement activation by binding of the negative complement regulator factor H. Mass spectrometric analysis revealed that wild type bacteria grown in vitro lacked any sialylated LPS, but became sialylated during infection. Sialic acid terminal residues of the core sugars of *H. influenzae* lipopolysaccharide (LPS) were shown to be a critical virulence factor in the pathogenesis of experimental otitis media in chinchillas since sialyltransferase-deficient (sialic acid devoid) mutants were profoundly attenuated [71]. Thus, *H. influenzae* exploits the host structural resources for its own protection. Strikingly, neuraminidase activity produced by *Streptococcus pneumoniae*, normally dwelling in the same ecological niche, the nasopharynx, seems to desialylate the protective layer of its microbial competitor *H. influenzae* and illustrates the tight competition on mucosal surfaces [72].

Evade Recognition and Cell Activation

It has recently become clear that epithelial cells at many anatomical sites of the body (similar to professional immune cells) express receptor molecules that recognize specific microbial structures, so called pathogen-associated molecular patterns (PAMPs) [73]. The family of toll-like receptors (TLRs) has gained substantial attention due to their role in cellular stimulation and systemic immune response activation [74]. However, their functional importance for the regulation of the epithelial host defense and other physiological responses is only beginning to be understood. TLRs are activated by conserved microbial structures that show little variation within whole groups of microbes. Examples are the lipopolysaccharide (LPS) that is found in all gram-negative bacteria, or lipoteichoic acid, a surface molecule of gram-positive bacteria. Thus, this system allows the detection and rough differentiation of microbial organisms with a limited set of receptors.

A straightforward evasion from TLR-mediated recognition could be to shield exposed ligands with a less immunostimulatory capsule or capsule-like structure. Indeed many pathogens such as *S. pneumoniae*, *H. influenzae*, and certain types of *E. coli* express a thick capsule. Another strategy, modification of recognized structures, is impaired by the fact that most TLR ligands such as lipopolysaccharide, teichoic acid, di- or tri-acylated lipopeptides, flagellin or hypomethylated DNA have essential functions in microbial physiology and modification is associated with significant reduction in viability and fitness.

Nevertheless, some microbial strategies to alter the recognized structures have been reported. For example, *S. typhimurium*, the most frequent causative agent of food-born enteritis in man, confers multiple alterations to the LPS structure to lower the proinflammatory potential [75]. Interestingly, such modifications of the LPS structure is controlled by the PhoP/PhoQ [75] and PmrA-PmrB two-component signal transduction systems [76] which are central modulators of virulence gene expression. The coordinated regulation of LPS structure modification and virulence illustrates the significant contribution of immune evasion in the process of microbial pathogenicity. Another striking example of the adaptation of the LPS structure to the host comes from the study of pathogenic *Yersiniae*. *Y. enterocolitica* and *Y. pseudotuberculosis* cause food- or waterborne enteritis and abdominal lymphadenopathy; *Y. pestis* is the causative agent of plague, a highly infectious and usually lethal systemic infection transmitted by flea bites. Whereas growth at 21°C induces the production of highly immunostimulatory LPS molecules, incubation of the bacterial cultures at 37°C changes the LPS biosynthesis to create tetra-acylated, non-stimulatory LPS molecules [77]. *Yersiniae* might therefore specifically alter the LPS structure under the conditions seen during infection of the warm-blooded mammal to limit TLR4 stimulation and host defense activation. Also *Bacteroides fragilis*, a common member of the intestinal flora produces LPS molecules with low immunostimulatory potential [78]. In addition to the low immunostimulatory potential of its LPS molecules, *H. pylori*, causing persistent infection of the gastric mucosa, also provides an example of the modification of another TLR ligand: recombinant flagellin from *H. pylori* exerted 1000-fold less stimulatory potential as compared to flagellin from *S. typhimurium.* [79, 80].

Few examples have been reported that indicate direct interference of microbial pathogens with TLR-mediated immune stimulation such as, for example, for the vaccinia virus [81]. The V-antigen (LcrV) from *Yersinia* exploits CD14 and TLR2 to induce the immunomodulatory cytokine IL10 [82]. Increased production of this inhibitory cytokine then in turn downregulates host defense and promotes bacterial infection. TLR2 deficient mice showed enhanced resistance to oral infection with *Y. enterocolitica*. However, the molecular base of this LcrV/TLR2 interaction remains to be elucidated. Another example of direct interference with TLR-mediated signaling might be obtained from studies of *Porphyromonas gingivalis*, a principle bacterium in adult periodontitis. It secretes an arginine-specific cysteine proteinase [gingivapain-R] that efficiently cleaves the coreceptor of TLR4, membranous CD14, on fibroblasts [83]. Since TLR4 largely depends on its coreceptor CD14 to efficiently recognize LPS, the lack of CD14 significantly diminishes bacteria-mediated cell activation. In addition, *P. gingivalis* produces a LPS molecule with low immunostimulatory potencial [84].

Also, microbial interference with cellular activation signals generated following receptor stimulation might represent a strategy to inhibit host inflammatory responses. Prominent examples are provided by enteropathogenic *Yersiniae* that encode for a so-called protein type III secretion system, that literary functions as molecular syringe injecting virulence proteins into the host cell cytoplasm [85]. One of these, YopH is a protein tyrosine phosphatase that targets p130^{Cas}, focal adhesion kinase and paxillin [86, 87] and prevents efficient actin polymerization needed for phagocytosis of the intoxicating bacteria. Furthermore, YopH desphoshorylation also impairs B- and T-cell function and cytokine release [88, 89]. Moreover, the YopP/YopJ effector proteins inhibit the IKKβ kinase and thereby mediate a general downregulation in host cell signal transduction by preventing nuclear translocation of transcription factor NF-κB [90]. Such countermeasures to the activation of host innate defense responses are not unique to *Yersiniae*, but evidently present among other bacteria. *Mycobacteriae*, for example, can inhibit MAP kinase pathways [91] and expresses virulence-associated tyrosine phosphatases [92]. Thus, one way to circumvent inflammatory reactions could be through an active downregulation of the proinflammatory signal transduction pathway.

Beside ligand alteration and interference with receptor signalling, tropism to selected anatomical sites that are devoid of TLR expression might represent a possible strategy to avoid host defense activation. In fact, TLR4 expression by intestinal epithelial cells has been shown to be restricted to crypt epithelial cells [93]. The luminal crypt milieu is characterized by a high concentration of antimicrobial peptides that might kill any approaching bacteria and quench the stimulatory potential of soluble LPS. In contrast, epithelial cells lining the intestinal villi should not respond to LPS exposure and might therefore represent a suitable site for colonization. However, as discussed above, the villus epithelium is characterized by a high cell turnover which might limit microbial colonization. Another interesting observation is the restriction of TLR5 expression to the basolateral site of epithelial cells [94]. Again, microbial colonization on the apical site would not result in immune effector activation. Compartmentalization of PRR expression might therefore represent an important mechanism to allow microbial colonization but preserve a functional immune recognition system.

Finally, although avoidance of immune recognition seems to be the most suitable way that pathogens might persue to successfully infect a host, at least some pathogenic organisms obviously seem to intentionally provoke proinflammatory responses to break the epithelial barrier and gain access to the subepithelial space. For example, *P. aeruginosa* an opportunistic pathogen that chronically colonizes the lung of patients with cystic fibrosis produces a highly stimulatory LPS molecule in vivo. Hepta-acylated Lipid A was detected on *P. aeruginosa* strains isolated from infected patients. This hepta-acylated LPS structure conferred efficient recognition by TLR4 and stimulated marked proinflammatory mediator secretion. In the context of chronic intraluminal colonization of the airways with production of large amounts of extracellular matrix, inflammation might represent a mechanism to enhance the supply of nutrients [95]. Similarly, *Shigella dysenteriae*, the causative agent of bacillary dysentery in humans, encodes two copies of the *msbB* gene, that is critically involved in the biosynthesis of immunostimulatory LPS [96]. Bacterial invasion of *S. dysenteriae* might be facilitated by the secretion of proinflammatory mediators that induce local leukocyte infiltration and disruption of the enteric mucosal layer. However, a recent study revealed increased bacterial invasion and tissue destruction following antibody-mediated blockade of CD14 [97]. Thus LPS-mediated immune activation might at the same time promote epithelial barrier disruption and serve the stimulation of antimicrobial factors to reduce the bacterial burden and protect from microbial spread.

Interestingly, mutations within the gene encoding another pattern recognition receptor molecule, NOD2 have recently been associated with inflammatory bowel disease [98–100]. NOD2 recognizes parts of the peptidoglycan structure common to Gram-negative and Gram-positive bacteria and is able to activate NF-κB. Strikingly, it resides within the cytoplasm and stimulation requires the presence of ligands within the cytoplasm. Since the described mutations revealed a loss of function phenotype, this surprising finding indicates a negative feedback-loop that is instrumental to maintain the physiological homeostasis in the healthy individual. In fact, the importance of the microbial flora for the clinical course of inflammatory bowel disease has long been established. Thus the balance between the microbial flora and the epithelial barrier in the intestine seems indeed to rely on recognition and adequate adaptation (including antimicrobial defense activation) rather than on ignorance and passive protection, as has previously been presumed.

Withstand Targeted Destruction

Epithelial cells sense approaching microorganisms and secret proinflammatory mediators to recruit professional immune cells and activate a systemic immune response. However, already the epithelial surfaces themselves provide effective antimicrobial defense mechanisms to eliminate invading microbes or at least retard the process of microbial infection. Given the rapid proliferation of many microbial organisms (with generation times as short as 20–30 min) early local host-response activation, independent from recruited professional immune cells, might be critical in many clinical circumstances. In fact, the typical signs of infection, reddening, swelling, and pain, which are induced by the recruitment of systemic immune effector cells of the immune system through vasodilatation and tissue infiltration of phagocytes, are rarely seen in most minor lesions experienced in everyday life. Thus, most microbial challenges of the outer and inner epithelial surfaces (microlesions, skin erosions, ragades) might be successfully defeated by intrinsic host defense mechanisms of the epithelial cell layer. A large variety of antimicrobial peptides protect the outer and inner body surfaces against environmental microbial pathogens. The high diversity and species specificity within this group of molecules might reflect the requirement to compete with the much faster evolution of microbial organisms [101]. In addition, high concentrations of these potent substances are produced to control the growth of the microbial flora and protect from infection.

The major group of these antimicrobial substances is the group of antimicrobial peptides, small amphiphilic molecules with high resistance against proteolytic degradation. A large variety of different substances such as β-defensins, cathelicidin, psoriasin, and RNAse7 are found within the epithelial layer of the skin and mucosal organs in large quantities [102]. In addition, polymorphonuclear cells as well as specialized epithelial cells, so called Paneth cells, located at the lower end of small intestinal crypts produce α-defensins. Peptide secretion into the intestinal crypt lumen has been calculated to lead to a local concentration of approximately 10 mg/ml [103, 104]. Although the exact mechanisms of microbial killing by antimicrobial peptides remain to be elucidated, interaction with the microbial membrane represents an early and critical step. In accordance, most microbial organisms that show decreased susceptibility to antimicrobial peptides demonstrate alterations in their surface structures. The fact that bacterial membranes are less susceptible to the action of antimicrobial peptides than artificial membranes further indicates an evolutionary pressure to withstand low concentrations of antimicrobial peptides [105].

Initial binding of cationic antimicrobial peptides occurs through attraction towards the negatively charged microbial surface. The study of *S. aureus*, a prominent causative agent of wound infections and abscess formation, has provided several mechanisms to reduce the susceptibility towards antimicrobial peptides, a feature that might significantly contribute to the pathogenic potential of this organism. Addition of *D*-alanine to teichoic acid, a major outer membrane constituent of all gram-positive bacteria, addition of *L*-lysine to membrane phospholipids, as well as amidation of *D*-glucose in the peptidoglycan coat all represent modifications aimed at reducing the net negative charge of the bacterial membrane and lead to reduced binding of antimicrobial peptides on the microbial cell membrane [106, 107]. Similarly, also Salmonella has been described to modify its outer surface structure by a number of modifications of the LPS molecule [108].

In addition, although antimicrobial peptides have been demonstrated to be highly resistant towards proteolytic cleavage, degradation of at least some linear antimicrobial peptides by PgtE from *Salmonella* and *E. coli* has been described [109, 110]. Also active transmembrane transport and excretion from the microbial cell via the MtrCDE transport system has been reported in *Neisseria* [111]. Finally, bacterial cleavage of extracellular matrix molecules that bind antimicrobial peptides with high affinity and inhibit their action has been described. Mechanisms to resist peptide-mediated killing can clearly be interpreted as virulence factors [112]. This was recently illustrated when the mechanism used by the virulence factor encoded by the *mig14* gene of *Salmonella* was identified: Mig14 mutants showed enhanced susceptibility to antimicrobial peptide-mediated killing [113].

Strikingly, although the described alterations lower the susceptibility to antimicrobial peptide-mediated killing, high levels of resistance have rarely been described. It might be associated with high metabolic costs and a significant reduction in fitness. Resistance of *S. pyogenes*, towards the action of CRAMP, the murine homolog of the human cathelicidine LL37, was only achieved by continuous exposure and associated with reduced growth in enriched culture medium [114]. Also, treatment with magainin, an antimicrobial peptide derived from the skin secretions of the frog, in clinical studies did not reveal any resistant bacteria.

Besides the expression of selected peptides, reactive oxygen and nitrogen intermediates are included in the repertoire of antimicrobial effector compounds. While the expression of reactive oxygen and nitrogen species is usually defined to selected types of cells, such as phagocytic cells or endothelial cells, bacterial gastrointestinal infections are associated with increased levels of luminal nitric oxide [115], which would imply that luminal bacteria should require means to protect themselves. Indeed, enzymes, such as catalases and super oxide dismutases, aimed at degrading reactive oxygen species are known to be expressed by enteric bacteria such as *E. coli* and *Salmonellae*. Likewise, the same bacteria contain genes (*norRVW*) evidently aimed at metabolizing NO [116, 117]. *Salmonella* also contains a NO inducible gene, *hmpA*, coding for a flavohemoglobin that protects the bacteria from NO in human macrophages [118].

Nevertheless, even with protective measures at hand, bacteria such as *E. coli* and *S. typhimurium* do not mount full resistance and show sensitivity to substances such as hydrogen peroxide and nitric oxide. Therefore, an additional strategy includes the means to repair the damage. Indeed, free radicals are known to induce for example the SOS response [119], whereas recombination-deficient *S. typhimurium* are sensitized to the macrophage oxidative burst [120]. Recently, mutations within genes coding for proteasome components of *Mycobacterium tuberculosis* were described to result in increased susceptibility to reactive nitrogen intermediates (RNI) [121]. Likewise, inhibition of the proteasome function was associated with restricted fitness in a murine infection model and increased sensitivity to RNI.

Active Penetration of the Epithelial Cell Barrier

Whereas protected from most extracellular host defense molecules, obligate intracellular microorganisms have to deal with other specific measures of the host to limit microbial spread. For example, many obligate intracellular bacteria are capable of inhibiting apoptosis to maintain their intracellular niche. This is also seen with *Chlamydophila pneumoniae* (previously termed *Chlamydia pneumoniae*), a common cause of community acquired pneumonia [122, 123]. However, for *C. pneumoniae* to be released, activation of a cell death process may be important. Interestingly, the bacterium encodes a protein capable of interacting with mammalian death receptors in vitro and in vivo. This protein, 'chlamydia protein associating with death domains' (CADD),

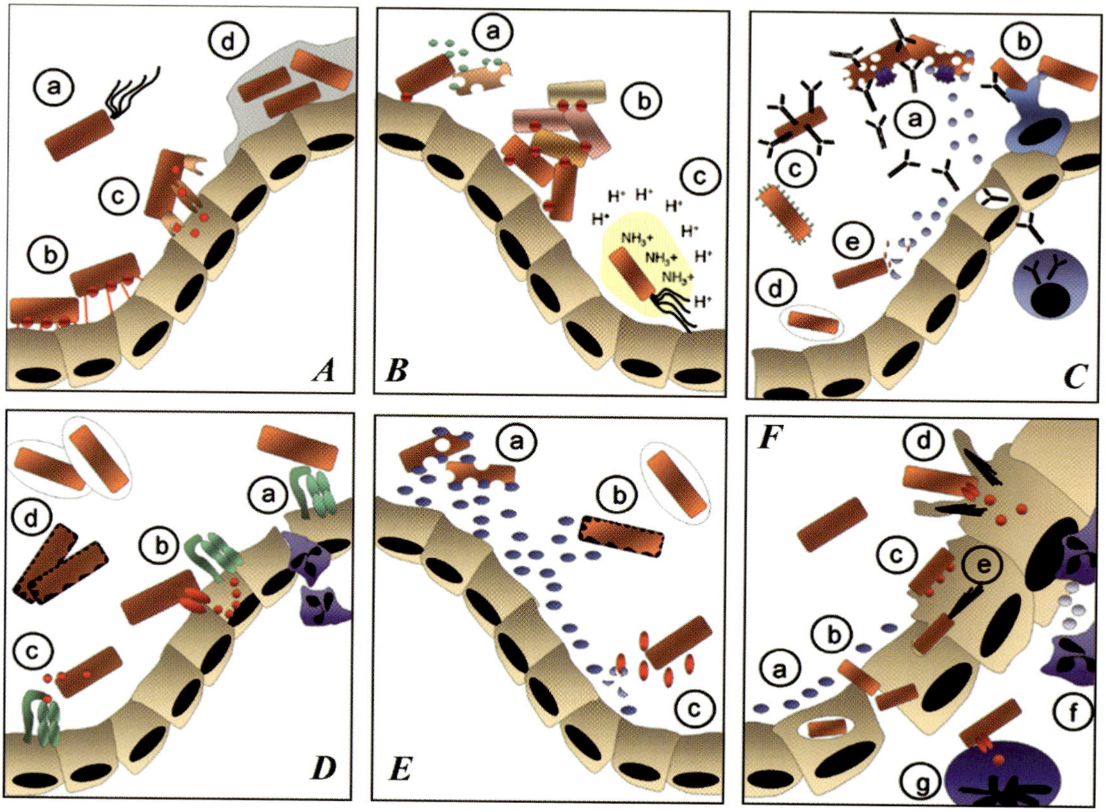

Fig. 1. Bacterial immune evasion mechanisms at epithelial surfaces. ***A*** Motility allows microbes to move towards suitable sites (a) and to attach via piliated or non-piliated adhesions (b), prerequisite for colonization. Enteropathogenic *E. coli* translocate their own receptor into the host cell (c). Alternatively attachment is facilitated by growth in a tight extracellular matrix or biofilm (d). ***B*** Bacterial members of the normal microbial flora compete through the production of bacteriocins (a) but also cooperate in host surface attachment (b). Specific pH buffering mechanisms exist to resist a low pH environment such as in the stomach (c). ***C*** Immunogloblins immobilize microbial organisms and together with complement factor deposition lead to microbial elimination through formation of the membrane attack complex (a) and enhanced receptor-mediated phagocytosis (b). Bacteria avoid antibody and complement deposition by recruitment of regulatory (inhibitory) complement factors or Fc-mediated antibody binding (c), surface shielding (d), and proteolytic degradation of complement factors (e). ***D*** Microbial organisms avoid cellular activation and phagocyte recruitment through pattern recognition receptors (PRR) (a) by interference with cellular signalling (b), modulation of receptor responses (c), or steric shielding and modification of recognized microbial structures (d). ***E*** Antimicrobial peptides protect epithelial surfaces from microbial colonization and infection (a). A decrease of susceptibility is facilitated by alterations in the bacterial surface structure (b), peptide export, and proteolytic peptide degradation (c). ***F*** Bacteria resist the epithelial

induces apoptosis in a variety of mammalian cells by engaging caspase activation, the hosts own apoptotic machinery [124].

Cellular invasion and intracellular persistence is also used by extracellularly growing bacteria to escape nutrient limitation and antibacterial activity. It was initially ascribed to a few pathogens, notably *Salmonellae*, *Shigellae* and *Yersiniae*, but it has later become evident that the ability of bacteria to enter host cells may be a more common trait. *H.pylori*, for example, has traditionally been regarded as a strictly extracellular bacterium, yet numerous studies have spotted the bacteria inside gastric epithelial cells [125]. Recently, detailed microscopic analyses have identified *H. pylori* to reside and survive within large intracellular vacuoles in a number of epithelial cell lines, and to actually cycle between extracellular and intravacuolar states [126].

Bacterial invasion involves the induction of phagocytosis or bacterial engulfment by non-phagocytic cells. Typically, the process is mastered by the action of selected bacterial proteins. Invasion of *Y. pseudotuberculosis* into mammalian cells relies on the bacterial Inv protein. Inv expression in laboratory *E. coli* is sufficient to enable this noninvasive bacterium to enter cultured mammalian cells [127]. Inv functions by binding to host cell β1 integrins, mammalian integrins have the ability to functionally connect to the actin cytoskeleton through activation of Rho GTPases and through recruitment of molecules mediating actin polymerization [128]. Thus, Inv evidently utilizes integrin-initiated actin polymerization to create the endocytotic event needed for bacterial internalization.

Invasion of both *Shigellae* and *Salmonellae* is likewise dependent on actin polymerization, but these pathogens apply an apparently more complex strategy for initiating endocytosis, and rely on type III secretion systems for initiating invasion [13, 129]. For *Shigella*, this invasion function is encoded by a virulence plasmid that also codes for Ipa effector proteins that activate Rho GTPases. These steer actin polymerization that is needed for bacterial uptake. Similarly, entry of *Salmonella* into non-phagocytic host cells depends on a protein type III secretion system coded for by the chromosomal pathogenicity island 1 (SPI1). Again it appears to be a similar strategy as seen in *Shigellae*; the *Salmonella* SopE effector protein is injected into the host cell, where it acts

defense by cellular invasion (a) through permeation of the epithelial lining, escape into the lamina propria, and basolateral invasion (b) or adhesion-mediated internalization (c). Activation of the actin cytoskeleton facilitates internalization (d), and intercellular spread (e). Cellular stimulation leads to the recruitment of phagocytes (f), which may be inhibited by the induction of apoptosis (g).

as a nucleotide exchange factor for Cdc42 and Rac-1. Additional effector proteins are involved in the modulation of actin polymerization, which subsequently results in bacterial uptake [130, 131].

One may ask why certain bacteria rely on complex machineries for their uptake, if the same process can be achieved by expressing a single virulence protein such as Inv of *Y. pseudotuberculosis*. One explanation could be that the effector proteins secreted into the host cell by *Shigellae* and *Salmonellae* are not only applied for inducing invasion, but also for inducing apoptosis and inflammation [132–135]. The *Shigella* IpaB effector protein necessary for invasion also mediates cleavage of caspase-1 in macrophages encountered at the site of invasion [132]. Thereby bacterial invasion becomes coupled to the release of IL-1β from the infected cells. Epithelial cells invaded by *Shigella* are also able to mount a proinflammatory response involving nuclear translocation of NF-κB and IL-8 production [136]. Combined, these effects cause neutrophil recruitment to the site of infection and subsequent amplification of the inflammatory reaction. Initially, this inflammation promotes bacterial infection by causing disintegration of the epithelial architecture [137], and blockage of neutrophil recruitment through application of anti-CD-18 antibodies were indeed reported to prevent bacterial invasion and tissue destruction [137, 138]. However, at the same time neutrophil engagement seems to prevent *Shigellae* from spreading from the lamina propria to the blood stream [139]. In essence, it seems that the bacterium provokes a strong inflammatory response in order to establish infection, only to be later eradicated through the responses elicited.

Like *Shigellae*, the *Salmonellae* SPI1 type III secretion machinery induces apoptosis and inflammatory responses. A strong activation of Cdc-42 and Rac-1 can in turn activate proinflammatory host signal transduction cascades [133]. Furthermore, one of the bacterial effector proteins injected, SipB, is capable of activating Caspase-1 and to subsequently cause release of IL-1β [140]. Thus, invasion of *Salmonella* is strongly coupled to the initiation of an inflammatory response. The importance of this response for the infection is well illustrated in a mouse model for systemic salmonellosis. Caspase-1 knockout mice are virtually resistant to a *per os* challenge of a virulent bacteria, but susceptible to an intraperitoneal challenge [141].

In conclusion, epithelial cells at many body sites are well-equipped to recognize and respond to microbial challenges and provide a wide spectrum of antimicrobial strategies. Nevertheless, microbial organisms have evolved an equally large number of mechanisms to avoid, circumvent, or modulate the host response as summarized in figure 1. The study of this interplay between microbial organisms and the host physiology and epithelial immune response will certainly further add to our understanding of this intimate relationship. Particularly the role of epithelial cells during microbial attachment, colonization, and penetration will allow identifying early steps during microbial pathogenesis. The initial microbe-host interaction might provide effective targets for therapeutic and prophylactic strategies to enhance resistance against microbial infections.

Acknowledgements

Owing to space limitations, many important and relevant original articles could only be cited indirectly through other reviews. The authors are supported by grants from the Swedish Research Council, the Swedish Foundation of Strategic Research, the Swedish Cancer Foundation, the Deutsche Forschungsgemeinschaft, and the EU 6th framework programme (project PREVIS).

References

1 Tancrede C: Role of human microflora in health and disease. Eur J Clin Microbiol Infect Dis 1992;11:1012–1015.

2 Jones GW, Rutter JM: Role of the K88 antigen in the pathogensis of neonatal diarrhea caused by *Escherichia coli* in pigs. Infect Immun 1972;6:918–927.

3 Duguid JP, Smith IW, Dempster G, Edmunds PN: Non-flagellar filamentous appendages (fimbriae) and haemagglutination activity in *Bacterium coli*. J Path Bact 1955;70:335–348.

4 Brinton CC Jr: The structure, function, synthesis and genetic control of bacterial pili and a molecular model for DNA and RNA transport in gram negative bacteria. Trans NY Acad Sci 1965;27:1003–1054.

5 Krogfelt KA: Bacterial adhesion: Genetics, biogenesis, and role in pathogenesis of fimbrial adhesins in *Escherichia coli*. Rev Infect Dis 1991;13:721–735.

6 Dodson KW, Pinkner JS, Rose T, Magnusson G, Hultgren SJ, Waksman G: Structural basis of the interaction of the pyelonephritogenic *Escherichia coli* adhesin to human kidney receptor. Cell 2001;105:733–743.

7 Korhonen TK, Parkkinen J, Hacker J, Finne J, Pere A, Rhen M and Holthöfer H: Binding of *Escherichia coli* S fimbriae to human kidney epithelium. Infect Immun 1986;54:322–327.

8 Wullt B, Bergsten G, Samuelsson M, Svanborg C: The role of P. fimbriae for *Escherichia coli* establishment and mucosal inflammation in the human urinary tract. Int J Antimicrob Agents 2002;19:522–538.

9 Roberts JA, Marklund BI, Ilver D, Haslam D, Kaack MB, Baskin G, Louis M, Möllby R, Winberg J, Normark S: The Gal(α1–4)Gal-specific tip adhesin of *Escherichia coli* P-fimbriae is needed for pyelonephritis to occur in the normal urinary tract. Proc Natl Acad Sci USA 1994;91: 11889–11893.

10 Winberg J, Möllby R, Bergström J, Karlsson KA, Leonardsson I, Milh MA, Teneberg S, Haslam D, Marklund BI, Normark S: The PapG adhesin at the tip of P-fimbriae provides *Escherichia coli* with a competitive edge in experimental bladder infections of cynomolgus monkeys. J Exp Med 1995;182:1695–1702.

11 Mulvey MA, Lopez-Boado YS, Wilson CL, Roth R, Parks WC, Heuser J, Hultgren SJ: Induction and evasion of host defenses by type 1-piliated uropathogenic *Escherichia coli*. Science 1998;282: 1494–1497. Erratum in: Science 1999;283:795.

12 Martinez JJ, Mulvey MA, Schilling JD, Pinkner JS, Hultgren SJ: Type 1 pilus-mediated bacterial invasion of bladder epithelial cells. EMBO J 2000;19:2803–2812.

13 Isberg RR, Voorhis DL, Falkow S: Identification of invasin: A protein that allows enteric bacteria to penetrate cultured mammalian cells. Cell 1987;50:769–778.

14 Gaillard JL, Berche P, Frehel C, Gouin E, Cossart P: Entry of *L. monocytogenes* into cells is mediated by internalin, a repeat protein reminiscent of surface antigens from gram-positive cocci. Cell 1991;65:1127–1141.

15 Vallance BA, Finlay BB: Exploitation of host cells by enteropathogenic *Escherichia coli*. Proc Natl Acad Sci USA 2000;97:8799–8806.

16 Celli J, Deng W, Finlay BB: Enteropathogenic *Escherichia coli* (EPEC) attachment to epithelial cells: Exploiting the host cell cytoskeleton from the outside. Cell Microbiol 2000;2:1–9.

17 Hansen-Wester I, Hensel M: *Salmonella* pathogenicity islands encoding type III secretion systems. Microbes Infect 2001;3:549–559.

18 Westerlund B, Korhonen TK: Bacterial proteins binding to the mammalian extracellular matrix. Mol Microbiol 1993;9:687–94.

19 Kuusela P: Fibronectin binds to *Staphylococcus aureus*. Nature 1978;276:718–720.

20 Patti JM, Allen BL, McGavin MJ, Hook M: MSCRAMM-mediated adherence of microorganisms to host tissues. Annu Rev Microbiol 1994;48:585–617.

21 Kukkonen M, Saarela S, Lähteenmäki K, Hynönen U, Westerlund-Wikström B, Rhen M, Korhonen TK: Identification of two laminin-binding fimbriae, the type-1 fimbria of *Salmonella enterica* serovar Typhimurium and the G fimbria of *Escherichia coli*, as plasminogen receptors. Infect Immun 1998;66:4965–4970.

22 Nickel JC, Ruseska I, Wright JB, Costerton JW: Tobramycin resistance of *Pseudomonas aeruginosa* cells growing as a biofilm on urinary catheter material. Antimicrob Agents Chemother 1985;27:619–624.

23 Marrie TJ, Costerton JW: Scanning and transmission electron microscopy of in situ bacterial colonization of intravenous and intraarterial catheters. J Clin Microbiol 1984;19:687–693.

24 Vuong C, Voyich JM, Fischer ER, Braughton KR, Whitney AR, DeLeo FR, Otto M: Polysaccharide intercellular adhesin (PIA) protects *Staphylococcus epidermidis* against major components of the human innate immune system. Cell Microbiol 2004;6:69–275.

25 Spoering AL, Lewis K: Biofilms and planktonic cells of *Pseudomonas aeruginosa* have similar resistance to killing by antimicrobials. J Bacteriol 2001;183:6746–6751.

26 Donlan RM, Costerton JW: Biofilms: Survival mechanisms of clinically relevant microorganisms. Clin Microbiol Rev 2002;15:167–193.

27 Anderson GG, Palermo JJ, Schilling JD, Roth R, Heuser J, Hultgren SJ: Intracellular bacterial biofilm-like pods in urinary tract infections. Science 2003;301:105–107.

28 Mulvey MA, Lopez-Boado YS, Wilson CL, Roth R, Parks WC, Heuser J, Hultgren SJ: Induction and evasion of host defenses by type 1-piliated uropathogenic *Escherichia coli*. Science 1998;282:1494–7.

29 Mulvey MA, Schilling JD, Martinez JJ, Hultgren SJ: Bad bugs and beleaguered bladders: Interplay between uropathogenic *Escherichia coli* and innate host defenses. Proc Natl Acad Sci USA 2000;97:8829–8835.

30 Grassme H, Kirschnek S, Riethmueller J, Riehle A, von Kurthy G, Lang F, Weller M, Gulbins E: CD95/CD95 ligand interactions on epithelial cells in host defense to *Pseudomonas aeruginosa*. Science 2000;290:527–530.

31 Flak TA, Goldman WE: Signalling and cellular specificity of airway nitric oxide production in pertussis. Cell Microbiol 1999;1:51–60.

32 Stoddart B, Wilcox MH: *Clostridium difficile*. Curr Opin Infect Dis 2002;5:513–518.

33 Ten Bruggencate SJ, Bovee-Oudenhoven IM, Lettink-Wissink ML, Katan MB, Van Der Meer R: Dietary fructo-oligosaccharides and inulin decrease resistance of rats to Salmonella: Protective role of calcium Gut 2004;53:530–535.

34 Buddington KK, Donahoo JB, Buddington RK: Dietary oligofructose and inulin protect mice from enteric and systemic pathogens and tumor inducers. J Nutr 2002;132:472–477.

35 Riley MA, Wertz JE: Bacteriocins: Evolution, ecology, and application. Annu Rev Microbiol 2002;56:117–137.

36 Kolenbrander PE, Andersen RN, Moore LV: Intrageneric coaggregation among strains of human oral bacteria: Potential role in primary colonization of the tooth surface. Appl Environ Microbiol 1990;56:3890–3894.
37 Kinder SA, Holt SC: Localization of the *Fusobacterium nucleatum* T18 adhesin activity mediating coaggregation with *Porphyromonas gingivalis* T22. J Bacteriol 1993;175:840–850.
38 Raymond KN, Dertz EA, Kim SS: Enterobactin: An archetype for microbial iron transport. Proc Natl Acad Sci USA 2003;100:3584–3588.
39 Cornelissen CN, Sparling PF: Iron piracy: Acquisition of transferrin-bound iron by bacterial pathogens. Mol Microbiol 1994;14:843–850.
40 Goetz DH, Holmes MA, Borregaard N, Bluhm ME, Raymond KN, Strong RK: The neutrophil lipocalin NGAL is a bacteriostatic agent that interferes with siderophore-mediated iron acquisition. Mol Cell 2002;10:1033–1043.
41 Nemeth E, Rivera S, Gabayan V, Keller C, Taudorf S, Pedersen BK, Ganz T: IL-6 mediates hypoferremia of inflammation by inducing the synthesis of the iron regulatory hormone hepcidin. J Clin Invest 2004;113:1271–1276.
42 Gangaidzo IT, Moyo VM, Mvundura E, Aggrey G, Murphree NL, Khumalo H, Saungweme T, Kasvosve I, Gomo ZA, Rouault T, Boelaert JR, Gordeuk VR: Association of pulmonary tuberculosis with increased dietary iron. J Infect Dis 2001;184:936–939.
43 Boskey ER, Telsch KM, Whaley KJ, Moench TR, Cone RA: Acid production by vaginal flora in vitro is consistent with the rate and extent of vaginal acidification. Infect Immun 1999;67:5170–5175.
44 McLean NW, McGroarty JA: Growth inhibition of metronidazole-susceptible and metronidazole-resistant strains of *Gardnerella vaginalis* by Lactobacilli in vitro. Appl Environ Microbiol 1996;62:1089–1092.
45 Cook GC: Infective gastroenteritis and its relationship to reduced gastric acidity. Scand J Gastroenterol 1985;111:17–23.
46 Banatvala N, Cramp A, Jones IR, Feldman RA: Salmonellosis in North Thames (East), UK: Associated risk factors. Epidemiol Infect 1999;122:201–207.
47 Nalin DR, Levine RJ, Levine MM, Hoover D, Bergquist E, McLaughlin J, Libonati J, Alam J, Hornick RB: Cholera, non-vibrio, and stomach acid. Lancet 1978;ii:856–859.
48 Foster JW, Moreno M: Inducible acid tolerance mechanisms in enteric bacteria. Novartis Found Symp 1999;221:55–69.
49 Bang IS, Audia JP, Park YK, Foster JW: Autoinduction of the *ompR* response regulator by acid shock and control of the *Salmonella enterica* acid tolerance response. Mol Microbiol 2002;44: 1235–1250.
50 Spector MP, Garcia del Portillo F, Bearson SM, Mahmud A, Magut M, Finaly BB, Dougan G, Foster JW, Pallen MJ: The *rpoS*-dependent starvation stress response locus stiA encodes a nitrate reductase (*narZYWV*) required for carbon-starvation-inducible thermotolerance and acid tolerance in *Salmonella typhimurium*. Microbiology 1999;145:3035–3045.
51 Saklani-Jusforgues H, Fontan E, Goossens PL: Effect of acid-adaptation on *Listeria monocytogenes* survival and translocation in a murine intragastric infection model. FEMS Microbial Lett 2000;193:155–159.
52 Stingl K, Altendorf K, Bakker EP: Acid survival of *Helicobacter pylori*: How does urease activity trigger cytoplasmic pH homeostasis? Trends Microbiol 2002;10:70–74.
53 Mollenhauer-Rektorschek M, Hanauer G, Sachs G, Melchers K: Expression of UreI is required for intragastric transit and colonization of gerbil gastric mucosa by Helicobacter pylori. Res Microbiol 2002;153:659–666.
54 Tombola F, Morbiato L, Del Giudice G, Rappuoli R, Zoratti M, Papini E: The *Helicobacter pylori* VacA toxin is a urea permease that promotes urea diffusion across epithelia. J Clin Invest 2001;108:929–937.
55 Debellis L, Papini E, Caroppo R, Montecucco C, Curci S: *Helicobacter pylori* cytotoxin VacA increases alkaline secretion in gastric epithelial cells. Am J Physiol 2001;281:G1440–G1448.
56 Eaton KA, Suerbaum S, Josenhans C, Krakowka S: Colonization of gnotobiotic piglets by *Helicobacter pylori* deficient in two flagellin genes. Infect Immun 1996;64:2445–2448.
57 Guentzel MN, Berry LJ: Motility as a virulence factor for *Vibrio cholerae.* Infect Immun 1975;11:890–897.

58 Fernie-King BA, Seilly DJ, Davies A, Lachmann PJ: Streptococcal inhibitor of complement inhibits two additional components of the mucosal innate immune system: Secretory leukocyte proteinase inhibitor and lysozyme. Infect Immun 2002;70:4908–4916.

59 Achtman M: Epidemic spread and antigenic variability of *Neisseria meningitidis*. Trends Microbiol 1995;3:186–192.

60 Quan CP, Berneman A, Pires R, Avrameas S, Bouvet JP: Natural polyreactive secretory immunoglobulin A autoantibodies as a possible barrier to infection in humans. Infect Immun 1997;65:3997–4004.

61 Mulks MH, Kornfeld SJ, Plaut AG: Specific proteolysis of human IgA by *Streptococcus pneumoniae* and *Haemophilus influenzae*. J Infect Dis 1980;141:450–456.

62 Medina E, Molinari G, Rohde M, Haase B, Chhatwal GS, Guzman CA: Fc-mediated nonspecific binding between fibronectin-binding protein I of *Streptococcus pyogenes* and human immunoglobulins. J Immunol 1999;163:3396–3402.

63 Forsgren A, Nordström K: Protein A from *Staphylococcus aureus*: The biological significance of its reaction with IgG. Ann NY Acad Sci 1974;236:252–266.

64 Pilz D, Vocke T, Heesemann J, Brade V: Mechanism of YadA-mediated serum resistance of *Yersinia enterocolitica* serotype O3. Infect Immun 1992;60:189–195.

65 Jarva H, Hellwage J, Jokiranta TS, Lehtinen MJ, Zipfel PF, Meri S: The group B streptococcal beta and pneumococcal Hic proteins are structurally related immune evasion molecules that bind the complement inhibitor factor H in an analogous fashion. J Immunol 2004;172:3111–3118.

66 Thern A, Stenberg L, Dahlbäck B, Lindahl G: Ig-binding surface proteins of *Streptococcus pyogenes* also bind human C4b-binding protein (C4BP), a regulatory component of the complement system. J Immunol 1995;154:375–386.

67 Åkesson P, Sjöholm AG, Björck L: Protein SIC, a novel extracellular protein of *Streptococcus pyogenes* interfering with complement function. J Biol Chem 1996;271:1081–1088.

68 Wexler DE, Nelson RD, Cleary PP: Human neutrophil chemotactic response to group A streptococci: Bacteria-mediated interference with complement-derived chemotactic factors. Infect Immun 1983;39:239–246.

69 Gilbert M, Watson DC, Cunningham AM, Jennings MP, Young NM, Wakarchuk WW: Cloning of the lipooligosaccharide α-2,3-sialyltransferase from the bacterial pathogens *Neisseria meningitidis* and *Neisseria gonorrhoeae*. J Biol Chem 1996;271:28271–28276.

70 Hood DW, Cox AD, Gilbert M, Makepeace K, Walsh S, Deadman ME, Cody A, Martin A, Mansson M, Schweda EK, Brisson JR, Richards JC, Moxon ER, Wakarchuk WW: Identification of a lipopolysaccharide α-2,3-sialyltransferase from *Haemophilus influenzae*. Mol Microbiol 2001;39:341–350.

71 Bouchet V, Hood DW, Li J, Brisson JR, Randle GA, Martin A, Li Z, Goldstein R, Schweda EK, Pelton SI, Richards JC, Moxon ER: Host-derived sialic acid is incorporated into *Haemophilus influenzae* lipopolysaccharide and is a major virulence factor in experimental otitis media. Proc Natl Acad Sci USA 2003;100:8898–8903.

72 Shakhnovich EA, King SJ, Weiser JN: Neuraminidase expressed by *Streptococcus pneumoniae* desialylates the lipopolysaccharide of *Neisseria meningitidis* and *Haemophilus influenzae*: A paradigm for interbacterial competition among pathogens of the human respiratory tract. Infect Immun 2002;70:7161–7164.

73 Bäckhed F, Hornef M: Toll-like receptor 4-mediated signaling by epithelial surfaces: Necessity or threat? Microbes Infect 2003;5:951–959.

74 Athman R, Philpott D: Innate immunity via Toll-like receptors and Nod proteins. Curr Opin Microbiol 2004;7:25–32.

75 Kawasaki K, Ernst RK, Miller SI: 3-O-deacylation of lipid A by PagL, a PhoP/PhoQ-regulated deacylase of *Salmonella typhimurium*, modulates signaling through Toll-like receptor 4. J Biol Chem 2004;279:20044–20048.

76 Gunn JS, Ryan SS, Van Velkinburgh JC, Ernst RK, Miller SI: Genetic and functional analyses of a PmrA-PmrB-regulated locus necessary for lipopolysaccharide modification, antimicrobial peptide resistance, and oral virulence of *Salmonella enterica* serovar Typhimrium. Infect Immun 2000;68:139–146.

77 Rebeil R, Ernst RK, Gowen BB, Miller SI, Hinnebusch BJ: Variation in lipid A structure in the pathogenic yersiniae. Mol Microbiol 2004;52:1363–1373.

78 Lindberg AA, Weintraub A, Zähringer U, Rietschel ET: Structure-activity relationships in lipopolysaccharides of *Bacteroides fragilis*. Rev Infect Dis 1990;12(suppl 2):S133–S141.

79 Bäckhed F, Rokbi B, Torstensson E, Zhao Y, Nilsson C, Seguin D, Normark S, Buchan AM, Richter-Dahlfors A: Gastric mucosal recognition of *Helicobacter pylori* is independent of Toll-like receptor 4. J Infect Dis 2003;187:829–836.

80 Gewirtz AT, Yu Y, Krishna US, Israel DA, Lyons SL, Peek RM Jr: *Helicobacter pylori* flagellin evades toll-like receptor 5-mediated innate immunity. J Infect Dis 2004;189:1914–1920.

81 Bowie A, Kiss-Toth E, Symons JA, Smith GL, Dower SK, O'Neill LA: A46R and A52R from vaccinia virus are antagonists of host IL-1 and toll-like receptor signaling. Proc Natl Acad Sci USA 2000;97:10162–10167.

82 Sing A, Rost D, Tvardovskaia N, Roggenkamp A, Wiedemann A, Kirschning CJ, Aepfelbacher M, Heesemann J: *Yersinia* V-antigen exploits toll-like receptor 2 and CD14 for interleukin 10-mediated immunosuppression. J Exp Med 2002;196:1017–1024.

83 Tada H, Sugawara S, Nemoto E, Takahashi N, Imamura T, Potempa J, Travis J, Shimauchi H, Takada H: Proteolysis of CD14 on human gingival fibroblasts by arginine-specific cysteine proteinases from *Porphyromonas gingivalis* leading to down-regulation of lipopolysaccharide-induced interleukin-8 production. Infect Immun 2002;70:3304–3307.

84 Ogawa T: Chemical structure of lipid A from *Porphyromonas* (*Bacteroides*) *gingivalis* lipopolysaccharide. FEBS Lett 1993;332:197–201.

85 Cornelis GR: *Yersinia* type III secretion: Send in the effectors. J Cell Biol 2002;158: 401–408.

86 Black DS, Bliska JB: Identification of p130^{cas} as a substrate of *Yersinia* YopH (Yop51), a bacterial protein tyrosine phosphatase that translocates into mammalian cells and targets focal adhesions. EMBO J 1997;16:2730–2744.

87 Persson C, Carballeira N, Wolf-Watz H, Fällman M: The PTPase YopH inhibits uptake of *Yersinia*, tyrosine phosphorylation of p130^{cas} and FAK, and the associated accumulation of these proteins in the peripheral focal adhesions. EMBO J 1997;16:2307–2318.

88 Yao T, Mecsas J, Healy JI, Falkow S, Chien Y: Suppression of T and B lymphocyte activation by a *Yersinia psedotuberculosis* virulence factor YopH. J Exp Med 1999;190:1343–1350.

89 Alonso A, Bottini N, Bruckner S, Rahmouni S, Williams S, Schoenberger SP, Mustelin T:Lck dephosphorylation at Tyr-394 and inhibition of T cell antigen receptor signaling by *Yersinia* phosphatase YopH. J Biol Chem 2004279:4922–4928.

90 Zhou L, Tan A, Hershenson MB: *Yersinia* YopJ inhibits pro-inflammatory molecule expression in human bronchial epithelial cells. Respir Physiol Neurobiol 2004;140:89–97.

91 Schorey JS, Cooper AM: Macrophage signaling upon mycobacterial infection: The MAP kinases lead the way. Cell Microbiol 2003;5:133–142.

92 Singh R, Rao V, Shakila H, Gupta R, Khera A, Dhar N, Singh A, Koul A, Singh Y, Naseema M, Narayanan PR, Paramasivan CN, Ramanathan VD, Tyagi AK: Disruption of *mptpB* impairs the ability of *Mycobacterium tuberculosis* to survive in guinea pigs. Mol Microbiol 2003; 50:751–762.

93 Ortega-Cava CF, Ishihara S, Rumi MA, Kawashima K, Ishimura N, Kazumori H, Udagawa J, Kadowaki Y, Kinoshita Y: Strategic compartmentalization of Toll-like receptor 4 in the mouse gut. J Immunol 2003;170:3977–3985.

94 Gewirtz AT, Navas TA, Lyons S, Godowski PJ, Madara JL: Cutting edge: Bacterial flagellin activates basolaterally expressed TLR5 to induce epithelial proinflammatory gene expression. J Immunol 2001;167:1882–1885.

95 Ernst RK, Hajjar AM, Tsai JH, Moskowitz SM, Wilson CB, Miller SI: *Pseudomonas aeruginosa* lipid A diversity and its recognition by Toll-like receptor 4. J Endotoxin Res 2003;9:395–400.

96 D'Hauteville H, Khan S, Maskell DJ, Kussak A, Weintraub A, Mathison J, Ulevitch RJ, Wuscher N, Parsot C, Sansonetti PJ: Two *msbB* genes encoding maximal acylation of lipid A are required for invasive *Shigella flexneri* to mediate inflammatory rupture and destruction of the intestinal epithelium. J Immunol 2002;168:5240–5251.

97 Wenneras C, Ave P, Huerre M, Arondel J, Ulevitch RJ, Mathison JC, Sansonetti P: Blockade of CD14 increases *Shigella*-mediated invasion and tissue destruction. J Immunol 2000;164: 3214–3221.
98 Ogura Y, Bonen DK, Inohara N, Nicolae DL, Chen FF, Ramos R, Britton H, Moran T, Karaliuskas R, Duerr RH, Achkar JP, Brant SR, Bayless TM, Kirschner BS, Hanauer SB, Nunez G, Cho JH: A frameshift mutation in NOD2 associated with susceptibility to Crohn's disease. Nature 2001;411:603–606.
99 Hugot JP, Chamaillard M, Zouali H, Lesage S, Cezard JP, Belaiche J, Almer S, Tysk C, O'Morain CA, Gassull M, Binder V, Finkel Y, Cortot A, Modigliani R, Laurent-Puig P, Gower-Rousseau C, Macry J, Colombel JF, Sahbatou M, Thomas G: Association of NOD2 leucine-rich repeat variants with susceptibility to Crohn's disease. Nature 2001;411:599–603.
100 Hampe J, Cuthbert A, Croucher PJ, Mirza MM, Mascheretti S, Fisher S, Frenzel H, King K, Hasselmeyer A, MacPherson AJ, Bridger S, van Deventer S, Forbes A, Nikolaus S, Lennard-Jones JE, Foelsch UR, Krawczak M, Lewis C, Schreiber S, Mathew CG: Association between insertion mutation in NOD2 gene and Crohn's disease in German and British populations. Lancet 2001;357:1925–1928.
101 Boman HG: Antibacterial peptides: Basic facts and emerging concepts. J Intern Med 2003;254: 197–215.
102 Schröder J-M: Epithelial antimicrobial peptides: Innate local host response elements. Cell Mol Life Sci 1999;56:32–46.
103 Zasloff M: Antimicrobial peptides of multicellular organisms. Nature 2002;415:389–395.
104 Ayabe T, Satchell DP, Wilson CL, Parks WC, Selsted ME, Ouellette AJ: Secretion of microbicidal alpha-defensins by intestinal Paneth cells in response to bacteria. Nat Immunol 2000;1: 113–118.
105 Matsuzaki K, Fukui M, Fujii N, Miyajima K: Interactions of an antimicrobial peptide, tachyplesin I, with lipid membranes. Biochim Biophys Acta 1991;1070:259–264.
106 Peschel A, Jack RW, Otto M, Collins LV, Staubitz P, Nicholson G, Kalbacher H, Nieuwenhuizen WF, Jung G, Tarkowski A, van Kessel KP, van Strijp JA: *Staphylococcus aureus* resistance to human defensins and evasion of neutrophil killing via the novel virulence factor MprF is based on modification of membrane lipids with L-lysine. J Exp Med 2001;193: 1067–1076.
107 Collins LV, Kristian SA, Weidenmaier C, Faigle M, Van Kessel KP, Van Strijp JA, Gotz F, Neumeister B, Peschel A: *Staphylococcus aureus* strains lacking D-alanine modifications of teichoic acids are highly susceptible to human neutrophil killing and are virulence attenuated in mice. J Infect Dis 2002;186:214–219.
108 Ernst RK, Guina T, Miller SI: *Salmonella typhimurium* outer membrane remodeling: Role in resistance to host innate immunity. Microbes Infect 2001;3:1327–1334.
109 Guina T, Yi EC, Wang H, Hackett M, Miller SI: A PhoP-regulated outer membrane protease of *Salmonella enterica* serovar Typhimurium promotes resistance to α-helical antimicrobial peptides. J Bacteriol 2000;182:4077–4086.
110 Stumpe S, Schmid R, Stephens DL, Georgiou G, Bakker EP: Identification of OmpT as the protease that hydrolyzes the antimicrobial peptide protamine before it enters growing cells of *Escherichia coli*. J Bacteriol 1998;180:4002–4006.
111 Shafer WM, Qu X-D, Waring AG, Lehrer RI: Modulation of *Neisseria gonorrhoeae* susceptibility to vertebrae antimicrobial peptides due to a member of the resistance/nodulation/division efflux pump family. Proc Natl Acad Sci USA 1998;97:1829–1833.
112 Groisman EA, Parra-Lopez C, Salcedo M, Lipps C, Heffron F: Resistance to host antimicrobial peptides is necessary for *Salmonella virulence*. Proc Natl Acad Sci USA 1992;89: 11939–943.
113 Brodsky IE, Ernst RK, Miller SI, Falkow S: *mig*-14 is a *Salmonella* gene that play a role in bacterial resistance to antimicrobial peptides. J Bacteriol 2002;184:3203–3213.
114 Nizet V, Ohtake T, Lauth X, Trowbridge J, Rudisill J, Dorschner RA, Pestonjamasp V, Piraino J, Huttner K, Gallo RL: Innate antimicrobial peptide protects the skin from invasive bacterial infection. Nature 2001;414:454–457.

115 Herulf M, Svennungsson B, Lagergren Å, Ljung T, Morcos E, Wiklund NP, Lundberg JON, Weitzberg E: Increased nitric oxide in infective gastroenteritis. J Infect Dis 1999;180: 542–545.
116 Gardner AM, Helmick RA, Gardner PR: Flavorubredoxin, an inducible catalyst for nitric oxide reduction and detoxification in *Escherichia coli*. J Biol Chem 2002;277:8172–8177.
117 Eriksson S, Lucchini S, Thompson A, Rhen M, Hinton JCD: Unravelling the biology of macrophage infection by gene expression profiling of intracellular *Salmonella enterica*. Mol Microbiol 2003;47:103–118.
118 Stevanin TM, Poole RK, Domencheaux EAG, Read RC: Flavohemoglobin Hmp protects *Salmonella enterica* serovar Typhimurium from nitric oxide related killing by human macrophages. Infect Immun 2002;70:4399–4405
119 Khil PP, Camerini-Otero RD: Over 1000 genes are involved in the DNA damage response of *Escherichia coli*. Mol Microbiol 2002;44:89–105.
120 Buchmeier NA, Lipps CJ, So MY, Heffron F: Recombination-deficient mutants of *Salmonella typhimurium* are avirulent and sensitive to the oxidative burst. Mol Microbiol 1993;7: 933–936.
121 Darwin KH, Ehrt S, Gutierrez-Ramos J-C, Weich N, Nathan CF: The proteasome of *Mycobacterium tuberculosis* is required for the resistance to nitric oxide. Science 2003;302:1963–1966.
122 Rajalingam K, Al-Younes H, Muller A, Meyer TF, Szczepek AJ, Rudel T: Epithelial cells infected with *C. pneumoniae* are resistant to apoptosis. Infect Immun 2001;69:7880–7888.
123 Fischer SF, Schwarz C, Vier J, Hacker G: Characterization of antiapoptotic activities of *Chlamydia pneumoniae* in human cells. Infect Immun 2001;69:7121–7129.
124 Stenner-Liewen F, Liewen H, Zapata JM, Pawlowski K, Godzik A, Reed JC: CADD, a *Chlamydia* protein that interacts with death receptors. J Biol Chem 2002;277:9633–9636.
125 Petersen AM, Krogfeldt KA: *Helicobacter pylori*: An invading microorganism? A review. FEMS Microbiol Lett 2003;25:117–126.
126 Amieva MR, Salama NR, Tompkins LS, Falkow S: *Helicobacter pylori* enter and survive within multivesicular vacuoles of epithelial cells. Cell Microbiol 2002;4:677–690.
127 Isberg RR, Voorhis DL, Falkow S: Identification of invasin: A protein that allows enteric bacteria to penetrate cultured mammalian cells. Cell 1987;50:769–778.
128 Giancotti FG, Ruoslahti E: Integrin signaling. Science 1999;285:1028–1032.
129 Sansonetti P: Host-pathogen interactions: The seduction of molecular cross-talk. Gut 2002;50: 1112–1118.
130 Zhou D, Galan J: *Salmonella* entry into host cells: The work in concert of type III secreted effector proteins. Microbes Infect 2001;3:1293–1298.
131 Galan JE, Zhou D: Striking balance: modulation of the actin cytoskeleton by *Salmonella*. Proc Natl Acad Sci USA 2000;97:8754–8761.
132 Sansonetti P: Microbes and microbial toxins: Paradigms for microbial-mucosal interactions. III. From symptoms to molecular pathogenesis. Am J Physiol 2001;280:G319–G323.
133 Sansonetti PJ, Phalipon A, Arondel J, Thirumalai K, Banerjee S, Akira S, Takeda K, Zychlinsky A: Caspase-1 activation of IL-1β and IL-18 are essential for *Shigella flexneri*-induced inflammation. Immunity 2000;12:581–590.
134 Hobbie S, Chen LM, Davis RJ, Galan JE: Involvement of mitogen-activated protein kinase pathway in the nuclear responses and cytokine production induced by *Salmonella typhimurium* in cultured intestinal epithelial cells. J Immunol 1997;159:5550–5559.
135 Lundberg U, Vinatzer U, Berdnik D, von Gabain A, Baccarini M: Growth phase-regulated induction of *Salmonella*-induced macrophage apoptosis correlates with transient expression of SPI-1 genes. J Bacteriol 1999;181:3433–3437.
136 Philpott DJ, Yamaoka S, Israel A, Sansonetti P: Invasive *Shigella flexneri* activates NF-κB through a lipopolysaccharide-dependent innate intracellular response and leads to IL-8 expression in epithelial cells. J Immunol 2000;165:903–914.
137 Perdomo JJ, Gounon P, Sansonetti PJ: Polymorphonuclear leukocyte transmigration promotes invasion of colonic epithelial cells by *Shigella flexneri*. J Clin Invest 1994;93: 633–643.

138 Perdomo JJ, Cavallion JM, Huerre M, Ohayon H, Gounon P, Sansonetti PJ: Acute inflammation causes epithelial invasion and mucosal destruction in experimental shigellosis. J Exp Med 1994; 180:1307–1319.
139 Sansonetti PJ, Arondel J, Huerre M, Harada A, Matsushima K: Interleukin-8 controls bacterial transepithelial translocation at a cost of epithelial destruction in experimental shigellosis. Infect Immun 1999;67:1471–1480.
140 Hersh D, Monack DM, Smith MR, Ghori N, Falkow S, Zychlinsky A: The *Salmonella* invasin SipB induces macrophage apoptosis by binding caspase-1. Proc Natl Acad Sci USA 1999;96: 2396–2401.
141 Monack DM, Hersh D, Ghori N, Bouley D, Zychlinsky A, Falkow S: *Salmonella* exploits caspase-1 to colonize Peyer's patches in a murine typhoid model. J Exp Med 2000;192:249–258.

Mathias Hornef, MD
Institute for Medical Microbiology and Hygiene
University of Freiburg, Hermann Herder Strasse 11, DE–79104 Freiburg (Germany)
Tel. +49 761 203 6547, Fax +49 761 203 6651, E-Mail mathias.hornef@uniklinik-freiburg.de

Kabelitz D, Schröder J-M (eds): Mechanisms of Epithelial Defense.
Chem Immunol Allergy. Basel, Karger, 2005, vol 86, pp 99–119

Recognition of Bacterial Products by Toll-Like Receptors

Holger Heine, Artur J. Ulmer

Department of Immunology and Cell Biology, Research Center Borstel,
Leibniz Center for Medicine and Biosciences, Borstel, Germany

Abstract

Over the past 5 years, our knowledge about how the immune system senses the microbial world has changed fundamentally. It has been known for decades that microbial products such as lipopolysaccharide or bacterial DNA have a profound activity on human cells. Whereas the molecular structure of many different pathogenic microbial compounds has been extensively studied and characterized, the molecular basis of their recognition by the immune system remained elusive for a long time. It was the late Charles Janeway who developed the idea of microbial structures forming pathogen-associated molecular patterns (PAMP) that would be recognized by pattern-recognition receptors [1]. Even if the notion of pattern recognition is challenged today, the discovery of the family of Toll receptors in species as diverse as *Drosophila* and humans, and the identification of their role in distinguishing molecules and structures that are common to microorganisms has led to a renewed appreciation of the innate immune system. This review focuses on the current knowledge about the different molecules that are recognized by Toll receptors in mammalian cells.

Mammalian Toll-Like Receptors and Their Ligands

Toll-like receptors (TLRs) have evolved as crucial system for alerting the host to the presence of numerous infectious agents, as well as for the release of endogenous signs of cell damage. In *Drosophila*, Toll activation is facilitated by different protease cascades that are able to process the same endogenous Toll activator spätzle, whereas the actual pattern recognition is performed by different pattern recognition receptors [see chapter by Imler and Bulet, this vol, pp 1–21]. This concept is fundamentally different from the mechanism by which mammalian TLRs are engaged: there is good evidence that mammalian TLRs can directly interact with their microbial ligands. This recognition of TLR ligands can take place either outside the cell at the cell membrane (TLR2 together with TLR1 or TLR6, TLR4, TLR5 and TLR11) or inside the cell (TLR3, TLR7, TLR8, and TLR9). Interestingly enough, this correlates somehow with the localization of their respective ligands.

Extracellular Recognition of TLR Ligands

TLR4

TLRs, which are known to recognize extracellular bacterial products on the cell membrane, are TLR1, TLR2, TLR4, TLR5, TLR6, and TLR11. The first microbial ligand, which was associated with TLR signaling, was lipopolysaccharide (LPS), the main component of the outer leaflet of the cell wall of gram-negative bacteria. At first, LPS was erroneously believed to signal through TLR2 [2, 3]. It was shown that HEK-293 cells, which do not express TLR2, translocate NFκB in response to LPS only after transfection with TLR2. However, this first definition of the signal transducing receptor for LPS already documented a serious problem which still hinders investigations in this field: the presence of hardly detectable or unknown contaminants in ligands that are believed to signal through TLRs. Later on, it became evident that it was not LPS itself, which induces TLR2-signaling, but a contamination within the commercial LPS-preparation, namely bacterial lipoprotein [4, 5].

By defining the gene which is responsible for the genetic defect in C3H/HeJ and C57BL/10ScCr mice causing unresponsiveness to LPS, the actual LPS-signaling receptor was identified to be TLR4. It was found that C3H/HeJ mice have a missense mutation in the third exon of the TLR4 gene, predicting the replacement of proline with histidine at position 712, whereas the C57BL/10ScCr mice are homozygous for a null mutation of TLR4. Finally, a LPS-nonresponsive phenotype was also found in TLR4 knockout mice [6]. In addition, work on natural [7] or generated [8] TLR2 gene-deficient animals also showed that TLR2 is not required for responses to bacterial LPS. For signaling, TLR4 needs the coexpression of an adaptor protein called MD-2 [9]. This protein consists of large disulfide-linked oligomers of dimeric subunits, and is produced and secreted by monocytes and dendritic cells. MD-2 binds closely to TLR4 and is essential for LPS responses [10–12]. In addition, as a soluble protein, MD-2 is able to interact directly with LPS [13] and enables LPS responses in HEK293 cells that express TLR4 [14].

Before TLR4 was identified as the signal-transducing LPS receptor, it had been found that serum contains a LPS-binding protein (LBP) which transfers LPS to CD14, a GPI-anchored cell-surface protein [15–17]. Moreover, the expression of CD14 on the cell membrane is necessary for specific binding of LPS [18]. In the absence of CD14, only unspecific binding is detectable, presumably due to intercalation of LPS into the cell membrane [19]. Nevertheless, genetic data clearly implicate physical contact of the lipid A portion of LPS with TLR4 [20, 21]. Furthermore, using modified radio labeled LPS molecules physical contact between LPS,TLR4 and MD-2 could be demonstrated, but only when CD14 is present [22]. Remarkable is the finding that the affinity of binding of LPS to TLR4/MD-2 complexes is about 10- to 20-fold higher than that with MD-2 or CD14 [23]. So far, results concerning the mode of action of MD-2 are conflicting: whereas Visintin et al. [24] found that MD-2 must be bound to TLR4 on the cell membrane before binding of LPS can occur, Kennedy et al. [25] identified the activating ligand of TLR4 as the MD-2/LPS complex. Just recently, the functional region of TLR4 required for association with MD-2 has been identified. It was shown that the amino-terminal subdomain with eight LRRs is the binding site for MD-2 [26].

LPS is an amphiphilic molecule consisting of a hydrophilic polysaccharide part and a covalently bound hydrophobic lipid component, termed lipid A. The polysaccharide part can be divided into two subdomains, the core region and the O-specific chain, composed of a sequence of repeating units of identical oligosaccharides [27]. In general, the lipid A component is composed of a phosphorylated β1,6-linked *D*-glucosamine disaccharide that carries up to six or seven acyl residues. Lipid A has been shown to represent the endotoxic principle of LPS, since the biological effects of LPS are reproduced by free lipid A [28]. Synthetic lipid A compounds were the basis for investigations on the structure-activity relationship of LPS and lipid A [29–32]. The minimal requirement for lipid A bioactivity, referred to the cytokine inducing capacity is a molecule having two *gluco*-configurated hexosamine residues, two phosphoryl groups, and six fatty acids as present in *E. coli* lipid A or compound 506 [33]. From a large number of data concerning the structure-activity relationship, one can conclude that the 'natural' form of the synthetic compound 506 represents the optimal configuration for monocyte-activating capacity. All chemically different substructures or derivatives of compound 506, either differing in the phosphorylation or in the acylation pattern of the hexosamine disaccharide are less or even not active in inducing cytokines in monocytes/macrophages.

The lack of the secondary fatty acids as in the tetraacylated lipid A precursor Ia (compound 406, also known as lipid IVa) renders this lipid A molecule completely inactive in human cells [34, 35]. These findings are further supported by the biologically inactive natural lipid A species of *Rhodobacter sphaeroides* and *Rhodobacter capsulatus*, which also differ in their acylation pattern from the *E. coli* lipid A [36, 37]. Both these natural substances and the synthetic lipid A precursor Ia showed unaffected binding activity and were therefore used as LPS antagonists for competition of the natural LPS during inflammation [35, 38]. A potent synthetic LPS antagonist in vitro as well as in vivo is compound E5531 which was synthesized based on the structure of the *Rhodobacter sphaeroides* lipid A [39, 40]. The conclusion that LPS and precursor Ia (or related compounds) compete downstream of the binding to CD14 [40, 41] has now been validated by the finding that these LPS antagonists compete for interaction with the LPS signal transducer TLR4 [20, 21].

In contrast to human monocytes, other structural requirements for the induction of monokine production by LPS are found in murine macrophages. For example, compound 406 possesses endotoxic activity in mice in vitro and in vivo [42–44], but is inactive in human cells [29, 34, 35, 45] and in various other primates [data not published]. Moreover, *R. sphaeroides* lipid A is a potent antagonist in human and mice, but acts as an LPS mimetic in cells of hamster origin [46]. This reaction has been found to be unrelated to the species of CD14 expressed [44]. Today, the reason for the different activity of lipid A has been determined: the species-specific activity of lipid A analogs depends on the species of TLR4 expressed [20, 21]. Lipid IVa binds to TLR4/MD-2 complexes in a similar strength as lipid A. Lipid IVa, however, does not induce human TLR4 oligomerization and inhibits lipid A induced human TLR4 oligomerization [47]. In addition, these authors find that the antagonistic activity of lipid IVa is defined by human MD-2 and not human TLR4.

LPS and lipid A are amphiphilic molecules and form supramolecular structures in aqueous media, air-water interfaces, and lipid bilayer membranes [48]. For LPS it has been shown that the molecular shape and the supramolecular conformation are important for its biological activity and the recognition by LPS binding molecules, e.g. TLR4 [49]. It was found that bioactive hexaacyl lipid A has a conical/concave shape, the cross-section of the hydrophobic region being larger than that of the hydrophilic region, whereas lipid IVa and other LPS antagonists have a cylindrical molecular shape. This indicates that a cylindrical shape seems to be a prerequisite for an antagonistic property, although this shape is not a sufficient attribute of antagonists (note that lipid IVa is only antagonistic in human but agonistic in murine cells).

CD14, TLR4 and MD-2 are the main players of the LPS receptor complex expressed on the cell membrane. However, substantial evidence indicates that additional extracellular molecules are also functional active components of the LPS-receptor complex. By fluorescence resonance energy transfer (FRET), Pfeiffer's group found a close proximity between CD14, CD55, CD47, CD64, and CD32 in unstimulated monocytes but a LPS-induced cluster comprising TLR4, CD14, CD55, CD64, CD11b/CD18 CD16, CD32, CD36, and CD81. Lipid IVa, however, is not able to induce such a clustering. A functional role of various of these molecules (besides CD14 and TLR4) in LPS signaling has been demonstrated for CD55 [50], CD11b/CD18 [51], and CD81 [52], e.g. similar to this observation, a cluster comprising TLR4, CD11b/CD18, CD55, CD81, hsp70, hsp90, GDF5, and CXCR4 has been observed only after induction with LPS or lipid A but not after stimulation with penta-acyl lipid A or lipid IVa [53]. These data indicate that the assembly of the LPS-receptor complex is not a fixed one in a given cell type but a very dynamic and variable one depending on the type of activation.

Besides LPS, a great number of further microbial and non-microbial compounds has been reported to signal through TLR4. A list of these compounds is given in table 1. For many of them, convincing evidence suggests that they indeed signal through TLR4 [54–56]. However, it is hard to envisage that all listed compounds, chemically so different, signal through the same receptor. Although in most cases the possible contamination with LPS was carefully considered, recent publications cast doubts about the normally used procedures to rule out LPS contamination as responsible for the detected TLR4 activity [57, 58]. Thus, one has to be extremely careful when investigating TLR4 ligands other than LPS. Highly purified or synthetic compounds and sophisticated chemical analyses are required to deal with this ongoing problem.

TLR2, TLR2/TLR1, and TLR2/TLR6

Similar to TLR4, there is also a wide list of microbial compounds which has been reported to signal through TLR2 (table 1). Prominent representatives are bacterial lipopeptides/lipoproteins (LP), lipoteichoic acid (LTA), and peptidoglycan (PG) [8, 59–63]. Firstly, LPS has falsely been described to signal through TLR2 due to contaminations of the LPS preparations with LP (see above) [4, 5]. However, for some atypical LPS, namely LPS from *L. interrogans* [64], *P. gingivalis* [65], *L. pneumophilia*, and *Rhizobium* species Sin-1 [66] signaling through TLR2 has been found. All these ligands are constituents of the bacterial cell wall, either of gram-negative and/or gram-positive bacteria.

Among the TLR2-dependent bacterial ligands, LP are primary candidates for analyzing the structural requirements in respect to their interaction with TLR2. First, these molecules are available from various different bacterial species expressing different molecular structures. Second, the chemical synthesis of LP analogues provides a vast variety of chemically defined molecules [67, 68]. Bacterial LP are di-O-acylated S-(2,3-dihydroxypropyl)-cysteinyl residues N-terminally coupled to distinct polypeptides. Examples include the MALP2 [69] and the LP from the N-terminus of the cytochrome subunit of the photoreaction center of *Rhodopseudomonas viridis* [70]. The S-(2,3-dihydroxypropyl)-cysteine may be acylated with a third fatty acid through an amide-linkage as it is the case for the LP from the cell wall of *E. coli* [71] and its synthetic analogue Pam3CSK4 [72].

Table 1. Currently described TLR receptor ligands*

TLR1 (via TLR2)
Bacterial products [114]
Tri-acylated lipoproteins [75, 77]
TLR2
Gram-negative bacteria [60–62,115–118]
LPS from
Leptospira interrogans [64]
Porphyromonas gingivalis LPS [65]
Legionella pneumophilia, Rhizobium [66]
Spirochetes [60, 119]
Mycobacteria [60, 120, 121]
Tri- and di-acylated bacterial lipoproteins [59, 60, 116, 122–124]
Lipoteichoic acid [61, 125]
Peptidoglycan [61, 62, 115]
Zymosan [116]
Mycobacterial lipoarabinomannan [126]
Other bacterial glycolipids [63, 127, 128]
BCG cell walls [129]
LPS-deficient mutant of *Neisseria meningitidis* [130, 131]
Modulin (PSM) [132]
Glycoproteins, *Prevotella intermedia* [133]
GPI anchors, *Trypanosoma cruzi* [134]
Soluble tuberculosis factor, PIM (mannosylated phosphatidyl inositol) [121, 135]
Heat shock protein 60 [136]
Necrotic cells [137]
Measles virus/protein H [138]
Bacterial fimbriae/synthetic peptides [139]
Polymannuronic acid [140]
Group B streptococci soluble factor [141]
Aspergillus fumigatus [142]
V antigen, *Yersinia enterocolitica* [143]
TLR3
Viral double-stranded RNA/poly I:C [99]
TLR4
Gram-negative bacteria [115, 144]
M. tuberculosis [121, 145]
BCG cell walls [129]
Lipopolysaccharide/lipid A [4, 6, 20, 21, 144, 146]
Acyclic lipid A agonist [147]
Other bacterial Glycolipids [63, 127, 128]

Table 1 (*continued*)

Taxol [54, 148, 149]
Respiratory syncytial virus/protein F [55, 150]
Heat shock protein 60 [136, 151]
Heat shock protein 70 [152]
Fibronectin A domain [153]
Fibrinogen [154]
Saturated fatty acids [155]
Cryptococcus [156]
Aspergillus fumigatus [157]
Lipoteichoic acid from treponema [128]
Polymannuronic acid [140]
Minmally modified LDL [158]
Surfactant protein-A [159]
β-Defensins [160]
Heparan sulfate [161]
Murine retroviruses [162]
TLR5
Flagellated bacteria, flagellin [92]
TLR6 (via TLR2)
di-acylated lipoprotein (MALP-2) [74, 76]
Peptidoglycan, G+ bacteria, zymosan [73]
Modulin (PSM) [132, 163]
Soluble tuberculosis factor [163]
Group B streptococci soluble [141]
Necrotic cells [137]
TLR7
Imiquimod, R848/resiquimod (imidazoquinolines) [102]
ssRNA [106–108]
TLR8 (human)
R848/resiquimod [103]
ssRNA [108]
TLR9
CpG bacterial DNA [109, 110, 164]
TLR11 (mice)
Uropathogenic bacteria [97]

* This table contains most of the currently described Toll receptor ligands. However, as already mentioned in the text, in particular for TLR2 and TLR4 ligands, one has to be aware of possible contaminations with bacterial lipoproteins, LTA or LPS, respectively.

Interestingly, unlike all other TLRs, which are apparently functional active as homodimers, TLR2 forms heterodimers either with TLR1 or TLR6 [73] to achieve specificity for different stimuli. Information about the structural factors determining the heterodimer usage of a given ligand became available through the generation of TLR2-, TLR6- and TLR1-deficient mice. The diacyl LP MALP2 requires TLR2 and TLR6 for signaling, whereas the triacylated synthetic compound Pam3CSK4 is able to induce activation of innate immune cells independently of TLR6 and mainly through TLR2/TLR1 heterodimers [74–77]. These findings and additional investigations using various di- and triacylated LP have led to the current model according to which triacylated LP signal through TLR2/TLR1 heterodimers, whereas diacylated LP signal through TLR2/TLR6 heterodimers [78]. However, we recently found that not only triacylated lipopeptides, but also diacylated lipopeptides like Pam2CSK4 and the C-terminally elongated MALP2 derivative Pam2CGNNDESNIS-FKEK-SK4 (MALP2-SK4) induce cell activation in a TLR6-independent manner. Our results, therefore, indicate that not only the acylation pattern but also the amino acid sequence of the peptide part of lipoproteins take part in the specificity of recognition by TLR2 heterodimers [79]. Similar to TLR4, there is also a species-specific recognition by TLR2 of distinct LP structures. It was found that HEK293 cells transfected with murine TLR2 translocate NF-κB in response to synthetic tri-lauroylated lipopeptide, whereas cells transfected with human TLR2 remain unresponsive [80].

LTA is a component of the cell wall of gram-positive bacteria and is discussed to be an immunostimulatory gram-positive counterpart of LPS [81]. Like LPS and LP, LTA is an amphiphilic molecule which forms micelles. It consists of a diacylglycerol-containing glycolipid anchor covalently coupled to a polymeric backbone structure with repeating units, which differ in different gram-positive species [82]. In *S. aureus*, the repeating units contain *D*-alanine and *D*-N-acetylglucosamine linked to a 1–3-polyglycerophosphate chain [83], whereas in LTA from *S. pneumoniae*, the polymeric structure consists of tetrasaccharide repeating units that contain phosphorylcholine and are linked to each other by ribitol phosphate [84]. A highly purified LTA, which has been isolated by a novel butanol extraction procedure, has been described to induce TLR2-signaling in a CD14 and LBP dependent manner [85]. It was shown that this butanol-extraction procedure prevents decomposition of LTA, which was observed after the conventionally used phenol-extraction procedure. This decomposition resulted in a loss of glycerophosphate units as well as *D*-alanine and α-*D*-N-acetylglucosamine (GlcNAc) substituents from the LTA backbone [86]. Finally, the synthesis of complete LTA with six glycerophosphate units carrying four alanine and one N-acetyl-glucosamine substituent was performed successfully. This synthetic LTA displayed the same potency to activate monocytes as native LTA and was also shown to be inactive in macrophages derived from TLR2 knock out mice but active in wild type mice [87]. Thus, LTA was confirmed as an active TLR2-dependent inflammatory compound of gram-positive bacteria.

Whereas signaling induced by LTA through TLR2 is well established, it remains to be confirmed whether TLR2/TLR1 or TLR2/TLR6 heterodimers are involved in this process. Han et al. [88] found that anti-TLR1 monoclonal antibodies reduced pneumococcal- and staphylococcal LTA induced TNF-release by human monocytes. However, in a transfection system with TLR2, TLR2/1, or TLR2/6 transfected HEK293 cells both TLR2/1 and TLR2/6 transfected HEK293 cells showed a synergistic effect to LTA in comparison to HEK293-TLR2 cells [89].

Beside bacterial LP and LTA, PG has often been used to investigate TLR2-dependend signaling and was thought to be a prototypic TLR2 ligand. PG is a constituent of the cell wall of both gram-negative and gram-positive bacteria. It possesses a polymeric structure whose core consists of β1,4-linked N-acetyl-muramic acid-N-acetyl-*D*-glucosamine. This glycosyl polymer is cross-linked by short peptides and forms a three-dimensional network (sacculus). PG was used either from commercial sources which is a cell wall preparation consisting of the empty sacculus or as soluble PG, released by bacteria upon treatment with β-lactam antibiotics blocking enzymatic cross-linking. Both cell wall PG and soluble PG have been described to mediate cell activation through TLR2 [8, 62]. However, recently PG has been described to stimulate innate immune cells through the intracellular PRRs Nod1 and Nod2. Whereas Nod1 detects PG of gram-negative bacteria containing the unusual diaminopimelic acid, Nod2 is a receptor for PG derived from gram-negative and gram-positive bacteria recognizing muramyl dipeptide as the minimal motif [90]. Consequently, the contribution of TLR2 in the recognition of PG was reinvestigated using highly purified PG preparations derived from different gram-positive and gram-negative bacteria. It was found that highly purified PG, after removal of LP- and LTA-contaminants, was no longer able to induce NFκB in HEK293-TLR2 cells and to induce IL-6 release in murine macrophages [89]. Therefore, it was concluded that not PG itself, but impurities of the PG preparations (most likely LTA or LP) are responsible for TLR2-signaling.

TLR10

The only orphan TLR, human TLR10, is most closely related to TLR1 and TLR6. Thus, it is tempting to speculate that also TLR10 plays a role in TLR2-dependent responses. However, a ligand or a functional role for TLR10 has yet to be identified. There is only one recent publication describing a genetic variation in human TLR10 that may be associated with asthma [91]. Interestingly, a mouse ortholog for TLR10 has not been found.

TLR5

The ligand for TLR5 is bacterial flagellin from both gram-positive and gram-negative bacteria [92]. Bacterial flagella filaments form a helical propeller which serves as a 'motor' of bacterial mobility. The filaments are assembled by 11 protofilaments composed of flagellin monomers [93]. It is only the flagellin monomer which displays inflammatory activity through TLR5 [94]. The conserved motifs within the flagellin, which are responsible for intermolecular contact between the monomers, have been shown found to be recognized by TLR5 [94, 95]. The physical interaction of flagellin and TLR5 has been confirmed by immunocoprecipitation from cell lysates as well as after binding to intact cells [94, 96]. It is likely that the flagellin binding site on TLR5 is positioned at the amino-acids 386–407 containing a leucine-rich repeat [96].

TLR11

TLR11 has just recently been identified and cloned. It was found that TLR11 is strongly expressed in the liver and the kidney of mice and that it serves as a sensor for uropathogenic *E. coli* but not for nonpathogenic strains [97]. In addition, TLR11 deficient mice were more susceptible to infection with uropathogenic *E. coli* than wild-type mice. Since TLR11 does not respond to any known TLR ligand, uropathogenic *E. coli* bacteria are believed to express or contain a TLR11-specific ligand. Interestingly, it seems that humans do not express full-length TLR11 due to a stop codon in the ORF. Currently, it is not known whether this is true for all humans or if this represents a genetic polymorphism similar to the one that has been described for TLR5 [98].

Intracellular Recognition of TLR Ligands

Many TLR ligands, such as LPS, LP and flagellin represent key molecules of the microbial outer membranes and are recognized by surface-bound TLRs such as TLR2, TLR4 and TLR5, respectively. However, other microbe-specific PAMPs like double-stranded (ds) or single-stranded (ss) RNA and bacterial DNA containg CpG motifs are not as easily accessible for the recognition by the innate immune system. It is now apparent that the recognition of these ligands by TLR receptors takes place in different intracellular compartments.

TLR3

TLR3 is the receptor for dsRNA [99], a molecular pattern that is generated by many viruses during their replication cycle. For most experiments with respect to TLR3, however, the synthetic dsRNA analogue polyinosinic-polycytidilic acid (poly I:C) has been used. Another dsRNA analogue, poly A:U, can stimulate cells through TLR3 as well, albeit with much reduced activity. The recognition of these dsRNA molecules is specific since neither single-stranded RNA such as poly C nor polydesoxyinosinic-polydesoxycytidylic acid (poly dI:dC) trigger cell stimulation through TLR3. TLR3 is located in as yet unidentified intracellular vesicles in which the signalling is initiated [100]. So far, no species-specific recognition of different dsRNA molecules has been observed for TLR3. Recently, endogenous mRNA released from or associated with necrotic cells has been described as a TLR3 ligand [101]. Although mRNA is considered to be a single-stranded RNA molecule, it has a certain degree of secondary structure which includes double-stranded regions. The hypothesis is that through recognition of these mRNA molecules, innate immune cells such as DCs acquire the ability to sense injured or damaged cells that were taken up by phagocytosis.

TLR7/TLR8

For TLR7 and TLR8, the first ligands described were synthetic antiviral compounds belonging to the group of imidazoquinolines [102, 103]. Two of these compounds, R-848 and imiquimod can activate via TLR7 in mice and TLR7 and TLR8 in human cells. In addition, certain guanosine analogues such as loxoribine have been shown to activate exclusively through TLR7 [104, 105], indicating that at least in humans, structural differences between TLR7 and TLR8 ligands exist. Finally, the natural ligands for human TLR8 and murine TLR7 were identified as single-stranded RNA molecules [106–108]. It is noteworthy to mention that basically all forms of ssRNA (even poly U) can signal via mouse TLR7, provided they are detected in endosomal compartments. Thus, rather the abnormal localization than a complicated structural motif induces TLR7 signalling. It remains to be seen what the natural ligands for human TLR7 and mouse TLR8 are. In humans we see some redundancy for TLR7 and TLR8 with respect to imidazoquinolines; however, mouse TLR8 has yet to be shown to be functionally active.

TLR9

TLR9 was the first of the intracellular TLRs of which the ligand was identified. TLR9 recognizes bacterial but not mammalian DNA [109]. It has been known for a long time that bacterial DNA is a potent immune activator of mammalian cells and that this activity depends on the presence of unmethylated CpG motifs. These motifs are also found in mammalian cells, but with a lower frequency and are usually methylated. In contrast to the recognition of nucleic acids by TLR3, TLR7 and 8, TLR9 shows a high degree of specificity. Conversion of the CpG motifs in any given sequence into GpC motifs renders these sequences completely inactive. Moreover, species-specific motifs exist and the recognition of these motifs is dependent on the species of TLR9 [110].

For example, the optimal murine CpG sequence is GACGTT, whereas GTCGTT is optimal for human TLR9. These findings provide strong evidence that CpG-DNA directly engages TLR9.

The intracellular TLRs TLR3, TLR7, TLR8 and TLR9 seem to be predominantly expressed within endosomes [104, 105, 111, 112] which depends on structural features of their cytoplasmic domains [113]. In addition, acidification of the vacuole as well as endosomal maturation is required for the activation of these TLRs. Thus, the innate immune system evolved a mechanism to detect and discriminate the intracellular existence of key viral and bacterial signatures at a cellular localization where host-derived molecules normally do not occur.

Conclusion

The field of innate immunity has gained enormous interest over the last 5 years. TLRs and their signalling pathways represent an ancient system of host defence. Much of the work over the last years has focused on defining ligands for the different TLRs (table 1). However, the molecular basis for this recognition is not known, raising the possibility that additional receptors are involved in this process. In addition, future work has to answer the question how the leucine-rich repeats of TLRs are able to distinguish such chemically diverse ligands.

References

1 Janeway CA Jr: Approaching the asymptote? Evolution and revolution in immunology. Cold Spring Harb Symp Quant Biol 1989;54:1–13.

2 Kirschning CJ, Wesche H, Merrill AT, Rothe M: Human Toll-like receptor 2 confers responsiveness to bacterial lipopolysaccharide. J Exp Med 1998;188:2091–2097.

3 Yang RB, Mark MR, Gray A, Huang A, Xie MH, Zhang M, Goddard A, Wood WI, Gurney AL, Godowski PJ: Toll-like receptor-2 mediates lipopolysaccharide-induced cellular signalling. Nature 1998;395:284–288.

4 Hirschfeld M, Ma Y, Weis JH, Vogel SN, Weis JJ: Cutting edge: Repurification of lipopolysaccharide eliminates signaling through both human and murine Toll-like receptor 2. J Immunol 2000;165:618–622.

5 Lee HK, Lee J, Tobias PS: Two lipoproteins extracted from *Escherichia coli* K-12 LCD25 lipopolysaccharide are the major components responsible for Toll-like receptor 2-mediated signaling. J Immunol 2002;168:4012–4017.

6 Hoshino K, Takeuchi O, Kawai T, Sanjo H, Ogawa T, Takeda Y, Takeda K, Akira S: Cutting edge: Toll-like receptor 4 (TLR4)-deficient mice are hyporesponsive to lipopolysaccharide: Evidence for TLR4 as the Lps gene product. J Immunol 1999;162:3749–3752.

7 Heine H, Kirschning CJ, Lien E, Monks BG, Rothe M, Golenbock DT: Cutting edge: Cells that carry A null allele for toll-like receptor 2 are capable of responding to endotoxin. J Immunol 1999;162:6971–6975.

8 Takeuchi O, Hoshino K, Kawai T, Sanjo H, Takada H, Ogawa T, Takeda K, Akira S: Differential roles of TLR2 and TLR4 in recognition of gram-negative and gram-positive bacterial cell wall components. Immunity 1999;11:443–451.
9 Shimazu R, Akashi S, Ogata H, Nagai Y, Fukudome K, Miyake K, Kimoto M: MD-2, a molecule that confers lipopolysaccharide responsiveness on Toll-like receptor 4. J Exp Med 1999;189: 1777–1782.
10 Akashi S, Shimazu R, Ogata H, Nagai Y, Takeda K, Kimoto M, Miyake K: Cutting edge: Cell surface expression and lipopolysaccharide signaling via the toll-like receptor 4-MD-2 complex on mouse peritoneal macrophages. J Immunol 2000;164:3471–3475.
11 Visintin A, Mazzoni A, Spitzer JA, Segal DM: Secreted MD-2 is a large polymeric protein that efficiently confers lipopolysaccharide sensitivity to Toll-like receptor 4. Proc Natl Acad Sci USA 2001;98:12156–12161.
12 Yang H, Young DW, Gusovsky F, Chow JC: Cellular events mediated by lipopolysaccharide-stimulated toll-like receptor 4. MD-2 is required for activation of mitogen-activated protein kinases and Elk-1. J Biol Chem 2000;275:20861–20866.
13 Viriyakosol S, Tobias PS, Kitchens RL, Kirkland TN: MD-2 binds to bacterial lipopolysaccharide. J Biol Chem 2001;276:38044–38051.
14 Schromm AB, Lien E, Henneke P, Chow JC, Yoshimura A, Heine H, Latz E, Monks BG, Schwartz DA, Miyake K, Golenbock DT: Molecular genetic analysis of an endotoxin non-responder mutant cell line: A point mutation in a conserved region of MD-2 abolishes endotoxin-induced signaling. J Exp Med 2001;194:79–88.
15 Aderem A, Ulevitch RJ: Toll-like receptors in the induction of the innate immune response. Nature 2000;406:782–787.
16 Schumann RR, Leong SR, Flaggs GW, Gray PW, Wright SD, Mathison JC, Tobias PS, Ulevitch RJ: Structure and function of lipopolysaccharide binding protein. Science 1990;249:1429–1431.
17 Wright SD, Ramos RA, Tobias PS, Ulevitch RJ, Mathison JC: CD14, a receptor for complexes of lipopolysaccharide (LPS) and LPS binding protein [see comments]. Science 1990;249: 1431–1433.
18 Kirkland TN, Finley F, Leturcq D, Moriarty A, Lee JD, Ulevitch RJ, Tobias PS: Analysis of lipopolysaccharide binding by CD14. J Biol Chem 1993;268:24818–24823.
19 Schromm AB, Brandenburg K, Rietschel ET, Flad HD, Carroll SF, Seydel U: Lipopolysaccharide-binding protein mediates CD14-independent intercalation of lipopolysaccharide into phospholipid membranes. FEBS Lett 1996;399:267–271.
20 Lien E, Means TK, Heine H, Yoshimura A, Kusumoto S, Fukase K, Fenton MJ, Oikawa M, Qureshi N, Monks B, Finberg RW, Ingalls RR, Golenbock DT: Toll-like receptor 4 imparts ligand-specific recognition of bacterial lipopolysaccharide. J Clin Invest 2000;105:497–504.
21 Poltorak A, Ricciardi-Castagnoli P, Citterio S, Beutler B: Physical contact between lipopolysaccharide and Toll-like receptor 4 revealed by genetic complementation. Proc Natl Acad Sci USA 2000;97:2163–2167.
22 da Silva CJ, Soldau K, Christen U, Tobias PS, Ulevitch RJ: Lipopolysaccharide is in close proximity to each of the proteins in its membrane receptor complex: Transfer from CD14 to TLR4 and MD-2. J Biol Chem 2001;276:21129–21135.
23 Akashi S, Saitoh S, Wakabayashi Y, Kikuchi T, Takamura N, Nagai Y, Kusumoto Y, Fukase K, Kusumoto S, Adachi Y, Kosugi A, Miyake K: Lipopolysaccharide interaction with cell surface Toll-like receptor 4-MD-2: Higher affinity than that with MD-2 or CD14. J Exp Med 2003; 198:1035–1042.
24 Visintin A, Latz E, Monks BG, Espevik T, Golenbock DT: Lysines 128 and 132 enable lipopolysaccharide binding to MD-2, leading to Toll-like receptor-4 aggregation and signal transduction. J Biol Chem 2003;278:48313–48320.
25 Kennedy MN, Mullen GE, Leifer CA, Lee C, Mazzoni A, Dileepan KN, Segal DM: A complex of soluble MD-2 and lipopolysaccharide serves as an activating ligand for Toll-like receptor 4. J Biol Chem 2004;279:34698–34704.
26 Fujimoto T, Yamazaki S, Eto-Kimura A, Takeshige K, Muta T: The amino-terminal region of toll-like receptor 4 is essential for binding to MD-2 and receptor translocation to the cell surface. J Biol Chem 2004;279:47431–47437.

27 Lüderitz O, Freudenberg M, Galanos C, Lehmann V, Rietschel ET, Shaw D: Lipopolysaccharides of gram-negative bacteria; in Razin S, Rottern S (eds): Current Topics in Membrane and Transport. New York, Academic Press, 1982, pp 79–151.

28 Galanos C, Lüderitz O, Rietschel ETh, Westphal O, Brade H, Brade L, Freudenberg MA, Schade UF, Imoto M, Yoshimura H, Kusumoto S, Shiba T: Synthetic and *Escherichia coli* free lipid A express identical endotoxic activities. Eur J Biochem 1985;148:1–5.

29 Feist W, Ulmer AJ, Musehold J, Brade H, Kusumoto S, Flad HD: Induction of tumor necrosis factor-alpha release by lipopolysaccharide and defined lipopolysaccharide partial structures. Immunobiology 1989;179:293–307.

30 Rietschel ET, Brade L, Brandenburg K, Flad HD, de Jong-Leuveninck J, Kawahara K, Lindner B, Loppnow H, Luderitz T, Schade U: Chemical structure and biologic activity of bacterial and synthetic lipid A. Rev Infect Dis 1987;9(suppl 5):S527–S536.

31 Rietschel ET, Brade L, Schade FU, Seydel U, Zahringer U, Kusumoto S, Brade H: Bacterial endotoxins: Properties and structure of biologically active domains; in Schrinner E, Richmont MH, Seibert G, Schwarz U (eds): Surface of Microorganisms and Their Interactions with the Mammalian Host. Weinheim, Verlag Chemie, 1988, pp 1–41.

32 Wang M-H, Chen Y-Q, Flad H-D, Baer HH, Feist W, Ulmer AJ: Inhibition of interleukin-6 release and T-cell proliferation by synthetic mirror pseudo cord factor analogues in human peripheral blood mononuclear cells. FEMS Immunol Med Microbiol 1993;6:53–62.

33 Rietschel ET, Kirikae T, Feist W, Loppnow H, Zabel P, Brade L, Ulmer AJ, Brade H, Seydel U, Zahringer U, Schlaak M, Flad HD, Schade FU: Molecular aspects of the chemistry and biology of endotoxin; in Sies H, Flohe L, Zimmer G (eds): Molecular Aspects of Inflammation (42. Colloquium Mosbach, 1991). Berlin, Springer, 1991, pp 207–231.

34 Loppnow H, Brade L, Brade H, Rietschel ET, Kusumoto S, Shiba T, Flad HD: Induction of human interleukin 1 by bacterial and synthetic lipid A. Eur J Immunol 1986;16:1263–1267.

35 Kovach NL, Yee E, Munford RS, Raetz CR, Harlan JM: Lipid IVA inhibits synthesis and release of tumor necrosis factor induced by lipopolysaccharide in human whole blood ex vivo. J Exp Med 1990;172:77–84.

36 Loppnow H, Libby P: Proliferating or interleukin 1-activated human vascular smooth muscle cells secrete copious interleukin 6. J Clin Invest 1990;85:731–738.

37 Qureshi N, Takayama K, Kurtz R: Diphosphoryl lipid A obtained from the nontoxic lipopolysaccharide of *Rhodopseudomonas sphaeroides* is an endotoxin antagonist in mice. Infect Immun 1991;59:441–444.

38 Wang M-H, Flad H-D, Feist W, Brade H, Kusumoto S, Rietschel ET, Ulmer AJ: Inhibition of endotoxin-induced interleukin-6 production by synthetic lipid A structures in human peripheral blood mononuclear cells. Infect Immun 1991;59:4655–4664.

39 Christ WJ, Asano O, Robidoux AL, Perez M, Wang Y, Dubuc GR, Gavin WE, Hawkins LD, McGuinness PD, Mullarkey MA: E5531, a pure endotoxin antagonist of high potency. Science 1995;268:80–83.

40 Kitchens RL, Munford RS: Enzymatically deacylated lipopolysaccharide (LPS) can antagonize LPS at multiple sites in the LPS recognition pathway. J Biol Chem 1995;270:9904–9910.

41 Kitchens RL, Ulevitch RJ, Munford RS: Lipopolysaccharide (LPS) partial structures inhibit responses to LPS in a human macrophage cell line without inhibiting LPS uptake by a CD14-mediated pathway. J Exp Med 1992;176:485–494.

42 Galanos C, Lehmann V, Luderitz O, Rietschel ET, Westphal O, Brade H, Brade L, Freudenberg MA, Hansen-Hagge T, Luderitz T: Endotoxic properties of chemically synthesized lipid A part structures: Comparison of synthetic lipid A precursor and synthetic analogues with biosynthetic lipid A precursor and free lipid A. Eur J Biochem 1984;140:221–227.

43 Kirikae T, Schade FU, Zahringer U, Kirikae F, Brade H, Kusumoto S, Kusama T, Rietschel ET: The significance of the hydrophilic backbone and the hydrophobic fatty acid regions of lipid A for macrophage binding and cytokine induction. FEMS Immunol Med Microbiol 1994; 8:13–26.

44 Delude RL, Savedra RJ, Zhao H, Thieringer R, Yamamoto S, Fenton MJ, Golenbock DT: CD14 enhances cellular responses to endotoxin without imparting ligand-specific recognition. Proc Natl Acad Sci USA 1995;92:9288–9292.

45 Wang M-H, Flad H-D, Feist W, Musehold J, Kusumoto S, Brade H, Gerdes J, Rietschel ETh, Ulmer AJ: Inhibition of endotoxin or lipid A-induced tumor necrosis factor production by synthetic lipid A partial structures in human peripheral blood mononuclear cells. Lymphokine Cytokine Res 1992;11:23–31.
46 Golenbock DT, Hampton RY, Qureshi N, Takayama K, Raetz CR: Lipid A-like molecules that antagonize the effects of endotoxins on human monocytes. J Biol Chem 1991;266: 19490–19498.
47 Saitoh S, Akashi S, Yamada T, Tanimura N, Kobayashi M, Konno K, Matsumoto F, Fukase K, Kusumoto S, Nagai Y, Kusumoto Y, Kosugi A, Miyake K: Lipid A antagonist, lipid IVa, is distinct from lipid A in interaction with Toll-like receptor 4 (TLR4)-MD-2 and ligand-induced TLR4 oligomerization. Int Immunol 2004;16:961–969.
48 Seydel U, Hawkins L, Schromm AB, Heine H, Scheel O, Koch MH, Brandenburg K: The generalized endotoxic principle. Eur J Immunol 2003;33:1586–1592.
49 Schromm AB, Brandenburg K, Loppnow H, Moran AP, Koch MH, Rietschel ET, Seydel U: Biological activities of lipopolysaccharides are determined by the shape of their lipid A portion. Eur J Biochem 2000;267:2008–2013.
50 Heine H, el Samalouti VT, Notzel C, Pfeiffer A, Lentschat A, Kusumoto S, Schmitz G, Hamann L, Ulmer AJ: CD55/decay accelerating factor is part of the lipopolysaccharide-induced receptor complex. Eur J Immunol 2003;33:1399–1408.
51 Ingalls RR, Golenbock DT: CD11c/CD18, a transmembrane signaling receptor for lipopolysaccharide. J Exp Med 1995;181:1473–1479.
52 Malzan A, Nötzel C, Orinska Z, Bulfone-Paus S, Brade L, Ulmer AJ, Heine H: CD81 deficient mice are protected from LPS-induced shock. J Endotox Res 2002;8:198.
53 Triantafilou M, Brandenburg K, Kusumoto S, Fukase K, Mackie A, Seydel U, Triantafilou K: Combinational clustering of receptors following stimulation by bacterial products determines lipopolysaccharide responses. Biochem J 2004;381:527–536.
54 Kawasaki K, Akashi S, Shimazu R, Yoshida T, Miyake K, Nishijima M: Mouse toll-like receptor 4.MD-2 complex mediates lipopolysaccharide-mimetic signal transduction by Taxol. J Biol Chem 2000;275:2251–2254.
55 Kurt-Jones EA, Popova L, Kwinn L, Haynes LM, Jones LP, Tripp RA, Walsh EE, Freeman MW, Golenbock DT, Anderson LJ, Finberg RW: Pattern recognition receptors TLR4 and CD14 mediate response to respiratory syncytial virus. Nat Immunol 2000;1:398–401.
56 Malley R, Henneke P, Morse SC, Cieslewicz MJ, Lipsitch M, Thompson CM, Kurt-Jones E, Paton JC, Wessels MR, Golenbock DT: Recognition of pneumolysin by Toll-like receptor 4 confers resistance to pneumococcal infection. Proc Natl Acad Sci USA 2003;100:1966–1971.
57 Gao B, Tsan MF: Endotoxin contamination in recombinant human Hsp70 preparation is responsible for the inducation of TNFalpha release by murine macrophages. J Biol Chem 2003;278: 22523–22529.
58 Tsan MF, Gao B: Endogenous ligands of Toll-like receptors. J Leukoc Biol 2004;76:514–519.
59 Hirschfeld M, Kirschning CJ, Schwandner R, Wesche H, Weis JH, Wooten RM, Weis JJ: Cutting edge: Inflammatory signaling by *Borrelia burgdorferi* lipoproteins is mediated by toll-like receptor 2. J Immunol 1999;163:2382–2386.
60 Lien E, Sellati TJ, Yoshimura A, Flo TH, Rawadi G, Finberg RW, Carroll JD, Espevik T, Ingalls RR, Radolf JD, Golenbock DT: Toll-like receptor 2 functions as a pattern recognition receptor for diverse bacterial products. J Biol Chem 1999;274:33419–33425.
61 Schwandner R, Dziarski R, Wesche H, Rothe M, Kirschning CJ: Peptidoglycan- and lipoteichoic acid-induced cell activation is mediated by toll-like receptor 2. J Biol Chem 1999;274: 17406–17409.
62 Yoshimura A, Lien E, Ingalls RR, Tuomanen E, Dziarski R, Golenbock D: Cutting edge: Recognition of Gram-positive bacterial cell wall components by the innate immune system occurs via Toll-like receptor 2. J Immunol 1999;163:1–5.
63 Opitz B, Schroder NW, Spreitzer I, Michelsen KS, Kirschning CJ, Hallatschek W, Zahringer U, Hartung T, Gobel UB, Schumann RR: Toll-like receptor (TLR)-2 mediates treponema glycolipid and lipoteichoic acid (LTA)-induced NF-kappa B translocation. J Biol Chem 2001;276: 22041–22047.

64 Werts C, Tapping RI, Mathison JC, Chuang TH, Kravchenko V, Saint G, Haake DA, Godowski PJ, Hayashi F, Ozinsky A, Underhill DM, Kirschning CJ, Wagner H, Aderem A, Tobias PS, Ulevitch RJ: Leptospiral lipopolysaccharide activates cells through a TLR2-dependent mechanism. Nat Immunol 2001;2:346–352.
65 Hirschfeld M, Weis JJ, Toshchakov V, Salkowski CA, Cody MJ, Ward DC, Qureshi N, Michalek SM, Vogel SN: Signaling by Toll-Like receptor 2 and 4 agonists results in differential gene expression in murine macrophages. Infect Immun 2001;69:1477–1482.
66 Girard R, Pedron T, Uematsu S, Balloy V, Chignard M, Akira S, Chaby R: Lipopolysaccharides from Legionella and Rhizobium stimulate mouse bone marrow granulocytes via Toll-like receptor 2. J Cell Sci 2003;116:293–302.
67 Spohn R, Buwitt-Beckmann U, Brock R, Jung G, Ulmer AJ, Wiesmuller KH: Synthetic lipopeptide adjuvants and Toll-like receptor 2–structure-activity relationships. Vaccine 2004;22:2494–2499.
68 Wiesmuller KH, Bessler WG, Jung G: Solid phase peptide synthesis of lipopeptide vaccines eliciting epitope-specific B-, T-helper and T-killer cell response. Int J Pept Protein Res 1992;40:255–260.
69 Mühlradt PF, Kiess M, Meyer H, Sussmuth R, Jung G: Isolation, structure elucidation, and synthesis of a macrophage stimulatory lipopeptide from Mycoplasma fermentans acting at picomolar concentration. J Exp Med 1997;185:1951–1958.
70 Metzger JW, Beck-Sickinger AG, Loleit M, Eckert M, Bessler WG, Jung G: Synthetic S-(2,3-dihydroxypropyl)-cysteinyl peptides derived from the N-terminus of the cytochrome subunit of the photoreaction centre of *Rhodopseudomonas viridis* enhance murine splenocyte proliferation. J Pept Sci 1995;1:184–190.
71 Braun V: Covalent lipoprotein from the outer membrane of *Escherichia coli*. Biochim Biophys Acta 1975;415:335–377.
72 Metzger J, Wiesmuller KH, Schaude R, Bessler WG, Jung G: Synthesis of novel immunologically active tripalmitoyl-S-glycerylcysteinyl lipopeptides as useful intermediates for immunogen preparations. Int J Pept Protein Res 1991;37:46–57.
73 Ozinsky A, Underhill DM, Fontenot JD, Hajjar AM, Smith KD, Wilson CB, Schroeder L, Aderem A: The repertoire for pattern recognition of pathogens by the innate immune system is defined by cooperation between toll-like receptors. Proc Natl Acad Sci USA 2000;97:13766–13771.
74 Takeuchi O, Kawai T, Muhlradt PF, Morr M, Radolf JD, Zychlinsky A, Takeda K, Akira S: Discrimination of bacterial lipoproteins by Toll-like receptor 6. Int Immunol 2001;13:933–940.
75 Alexopoulou L, Thomas V, Schnare M, Lobet Y, Anguita J, Schoen RT, Medzhitov R, Fikrig E, Flavell RA: Hyporesponsiveness to vaccination with *Borrelia burgdorferi* OspA in humans and in TLR1- and TLR2-deficient mice. Nat Med 2002;8:878–884.
76 Morr M, Takeuchi O, Akira S, Simon MM, Muhlradt PF: Differential recognition of structural details of bacterial lipopeptides by toll-like receptors. Eur J Immunol 2002;32:3337–3347.
77 Takeuchi O, Sato S, Horiuchi T, Hoshino K, Takeda K, Dong Z, Modlin RL, Akira S: Cutting edge: Role of Toll-like receptor 1 in mediating immune response to microbial lipoproteins. J Immunol 2002;169:10–14.
78 Akira S: Mammalian Toll-like receptors. Curr Opin Immunol 2003;15:5–11 (published erratum Curr Opin Immunol 2003;15:238).
79 Buwitt-Beckmann U, Heine H, Wiesmuller KH, Jung G, Brock R, Akira S, Ulmer AJ: Toll-like receptor 6-independent signaling of diacylated lipopeptides. Eur J Immunol 2005;35:282–289.
80 Grabiec A, Meng G, Fichte S, Bessler W, Wagner H, Kirschning CJ: Human but not murine toll-like receptor 2 discriminates between tri-palmitoylated and tri-lauroylated peptides. J Biol Chem 2004;279:48004–48012.
81 Bhakdi S, Klonisch T, Nuber P, Fischer W: Stimulation of monokine production by lipoteichoic acids. Infect Immun 1991;59:4614–4620.
82 Greenberg JW, Fischer W, Joiner KA: Influence of lipoteichoic acid structure on recognition by the macrophage scavenger receptor. Infect Immun 1996;64:3318–3325.
83 Fischer W, Mannsfeld T, Hagen G: On the basic structure of poly(glycerophosphate) lipoteichoic acids. Biochem Cell Biol 1990;68:33–43.
84 Behr T, Fischer W, Peter-Katalinic J, Egge H: The structure of pneumococcal lipoteichoic acid. Improved preparation, chemical and mass spectrometric studies. Eur J Biochem 1992;207:1063–1075.

85 Schroder NW, Morath S, Alexander C, Hamann L, Hartung T, Zahringer U, Gobel UB, Weber JR, Schumann RR: Lipoteichoic acid (LTA) of *Streptococcus pneumoniae* and *Staphylococcus aureus* activates immune cells via Toll-like receptor (TLR)-2, lipopolysaccharide-binding protein (LBP), and CD14, whereas TLR-4 and MD-2 are not involved. J Biol Chem 2003;278:15587–15594.
86 Morath S, Geyer A, Spreitzer I, Hermann C, Hartung T: Structural decomposition and heterogeneity of commercial lipoteichoic acid preparations. Infect Immun 2002;70:938–944.
87 Deininger S, Stadelmaier A, von Aulock S, Morath S, Schmidt RR, Hartung T: Definition of structural prerequisites for lipoteichoic acid-inducible cytokine induction by synthetic derivatives. J Immunol 2003;170:4134–4138.
88 Han SH, Kim JH, Martin M, Michalek SM, Nahm MH: Pneumococcal lipoteichoic acid (LTA) is not as potent as staphylococcal LTA in stimulating Toll-like receptor 2. Infect Immun 2003;71:5541–5548.
89 Travassos LH, Girardin SE, Philpott DJ, Blanot D, Nahori MA, Werts C, Boneca IG: Toll-like receptor 2-dependent bacterial sensing does not occur via peptidoglycan recognition. EMBO Rep 2004;5:1000–1006.
90 Girardin SE, Philpott DJ: Mini-review: The role of peptidoglycan recognition in innate immunity. Eur J Immunol 2004;34:1777–1782.
91 Lazarus R, Raby BA, Lange C, Silverman EK, Kwiatkowski DJ, Vercelli D, Klimecki WJ, Martinez FD, Weiss ST: TOLL-like receptor 10 genetic variation is associated with asthma in two independent samples. Am J Respir Crit Care Med 2004;170:594–600.
92 Hayashi F, Smith KD, Ozinsky A, Hawn TR, Yi EC, Goodlett DR, Eng JK, Akira S, Underhill DM, Aderem A: The innate immune response to bacterial flagellin is mediated by Toll-like receptor 5. Nature 2001;410:1099–1103.
93 Samatey FA, Imada K, Nagashima S, Vonderviszt F, Kumasaka T, Yamamoto M, Namba K: Structure of the bacterial flagellar protofilament and implications for a switch for supercoiling. Nature 2001;410:331–337.
94 Smith KD, Andersen-Nissen E, Hayashi F, Strobe K, Bergman MA, Barrett SL, Cookson BT, Aderem A: Toll-like receptor 5 recognizes a conserved site on flagellin required for protofilament formation and bacterial motility. Nat Immunol 2003;4:1247–1253.
95 Donnelly MA, Steiner TS: Two nonadjacent regions in enteroaggregative *Escherichia coli* flagellin are required for activation of toll-like receptor 5. J Biol Chem 2002;277:40456–40461.
96 Mizel SB, West AP, Hantgan RR: Identification of a sequence in human toll-like receptor 5 required for the binding of gram-negative flagellin. J Biol Chem 2003;278:23624–23629.
97 Zhang D, Zhang G, Hayden MS, Greenblatt MB, Bussey C, Flavell RA, Ghosh S: A toll-like receptor that prevents infection by uropathogenic bacteria. Science 2004;303:1522–1526.
98 Hawn TR, Verbon A, Lettinga KD, Zhao LP, Li SS, Laws RJ, Skerrett SJ, Beutler B, Schroeder L, Nachman A, Ozinsky A, Smith KD, Aderem A: A common dominant TLR5 stop codon polymorphism abolishes flagellin signaling and is associated with susceptibility to legionnaires' disease. J Exp Med 2003;198:1563–1572.
99 Alexopoulou L, Holt AC, Medzhitov R, Flavell RA: Recognition of double-stranded RNA and activation of NF-kappaB by Toll-like receptor 3. Nature 2001;413:732–738.
100 Funami K, Matsumoto M, Oshiumi H, Akazawa T, Yamamoto A, Seya T: The cytoplasmic 'linker region' in Toll-like receptor 3 controls receptor localization and signaling. Int Immunol 2004;16:1143–1154.
101 Kariko K, Ni H, Capodici J, Lamphier M, Weissman D: mRNA is an endogenous ligand for Toll-like receptor 3. J Biol Chem 2004;279:12542–12550.
102 Hemmi H, Kaisho T, Takeuchi O, Sato S, Sanjo H, Hoshino K, Horiuchi T, Tomizawa H, Takeda K, Akira S: Small anti-viral compounds activate immune cells via the TLR7 MyD88-dependent signaling pathway. Nat Immunol 2002;3:196–200.
103 Jurk M, Heil F, Vollmer J, Schetter C, Krieg AM, Wagner H, Lipford G, Bauer S: Human TLR7 or TLR8 independently confer responsiveness to the antiviral compound R-848. Nat Immunol 2002;3:499.
104 Lee J, Chuang TH, Redecke V, She L, Pitha PM, Carson DA, Raz E, Cottam HB: Molecular basis for the immunostimulatory activity of guanine nucleoside analogs: Activation of Toll-like receptor 7. Proc Natl Acad Sci USA 2003;100:6646–6651.

105 Heil F, Ahmad-Nejad P, Hemmi H, Hochrein H, Ampenberger F, Gellert T, Dietrich H, Lipford G, Takeda K, Akira S, Wagner H, Bauer S: The Toll-like receptor 7 (TLR7)-specific stimulus loxoribine uncovers a strong relationship within the TLR7, 8 and 9 subfamily. Eur J Immunol 2003;33:2987–2997.
106 Lund JM, Alexopoulou L, Sato A, Karow M, Adams NC, Gale NW, Iwasaki A, Flavell RA: Recognition of single-stranded RNA viruses by Toll-like receptor 7. Proc Natl Acad Sci USA 2004;101:5598–5603.
107 Diebold SS, Kaisho T, Hemmi H, Akira S, Reis e Sousa: Innate antiviral responses by means of TLR7-mediated recognition of single-stranded RNA. Science 2004;303:1529–1531.
108 Heil F, Hemmi H, Hochrein H, Ampenberger F, Kirschning C, Akira S, Lipford G, Wagner H, Bauer S: Species-specific recognition of single-stranded RNA via toll-like receptor 7 and 8. Science 2004;303:1526–1529.
109 Hemmi H, Takeuchi O, Kawai T, Kaisho T, Sato S, Sanjo H, Matsumoto M, Hoshino K, Wagner H, Takeda K, Akira S: A Toll-like receptor recognizes bacterial DNA. Nature 2000;408: 740–745.
110 Bauer S, Kirschning CJ, Hacker H, Redecke V, Hausmann S, Akira S, Wagner H, Lipford GB: Human TLR9 confers responsiveness to bacterial DNA via species-specific CpG motif recognition. Proc Natl Acad Sci USA 2001;98:9237–9242.
111 Ahmad-Nejad P, Hacker H, Rutz M, Bauer S, Vabulas RM, Wagner H: Bacterial CpG-DNA and lipopolysaccharides activate Toll-like receptors at distinct cellular compartments. Eur J Immunol 2002;32:1958–1968.
112 Latz E, Schoenemeyer A, Visintin A, Fitzgerald KA, Monks BG, Knetter CF, Lien E, Nilsen NJ, Espevik T, Golenbock DT: TLR9 signals after translocating from the ER to CpG DNA in the lysosome. Nat Immunol 2004;5:190–198.
113 Nishiya T, DeFranco AL: Ligand-regulated chimeric receptor approach reveals distinctive subcellular localization and signaling properties of the Toll-like receptors. J Biol Chem 2004;279: 19008–19017.
114 Wyllie DH, Kiss-Toth E, Visintin A, Smith SC, Boussouf S, Segal DM, Duff GW, Dower SK: Evidence for an accessory protein function for Toll-like receptor 1 in anti-bacterial responses. J Immunol 2000;165:7125–7132.
115 Takeuchi O, Hoshino K, Kawai T, Sanjo H, Takada H, Ogawa T, Takeda K, Akira S: Differential roles of TLR2 and TLR4 in recognition of gram-negative and gram-positive bacterial cell wall components. Immunity 1999;11:443–451.
116 Underhill DM, Ozinsky A, Hajjar AM, Stevens A, Wilson CB, Bassetti M, Aderem A: The Toll-like receptor 2 is recruited to macrophage phagosomes and discriminates between pathogens [see comments]. Nature 1999;401:811–815.
117 Flo TH, Halaas O, Lien E, Ryan L, Teti G, Golenbock DT, Sundan A, Espevik T: Human toll-like receptor 2 mediates monocyte activation by *Listeria monocytogenes*, but not by group B streptococci or lipopolysaccharide. J Immunol 2000;164:2064–2069.
118 Takeuchi O, Hoshino K, Akira S: Cutting edge: TLR2-deficient and MyD88-deficient mice are highly susceptible to *Staphylococcus aureus* infection. J Immunol 2000;165:5392–5396.
119 Wooten RM, Ma Y, Yoder RA, Brown JP, Weis JH, Zachary JF, Kirschning CJ, Weis JJ: Toll-like receptor 2 is required for innate, but not acquired, host defense to *Borrelia burgdorferi*. J Immunol 2002;168:348–355.
120 Underhill DM, Ozinsky A, Smith KD, Aderem A: Toll-like receptor-2 mediates mycobacteria-induced proinflammatory signaling in macrophages. Proc Natl Acad Sci USA 1999;96: 14459–14463.
121 Means TK, Wang S, Lien E, Yoshimura A, Golenbock DT, Fenton MJ: Human Toll-like receptors mediate cellular activation by *Mycobacterium tuberculosis*. J Immunol 1999;163: 3920–3927.
122 Brightbill HD, Libraty DH, Krutzik SR, Yang RB, Belisle JT, Bleharski JR, Maitland M, Norgard MV, Plevy SE, Smale ST, Brennan PJ, Bloom BR, Godowski PJ, Modlin RL: Host defense mechanisms triggered by microbial lipoproteins through toll-like receptors. Science 1999;285: 732–736.

123 Aliprantis AO, Yang RB, Mark MR, Suggett S, Devaux B, Radolf JD, Klimpel GR, Godowski P, Zychlinsky A: Cell activation and apoptosis by bacterial lipoproteins through toll-like receptor-2. Science 1999;285:736–739.

124 Takeuchi O, Kaufmann A, Grote K, Kawai T, Hoshino K, Morr M, Muhlradt PF, Akira S: Cutting edge: Preferentially the R-stereoisomer of the mycoplasmal lipopeptide macrophage-activating lipopeptide-2 activates immune cells through a Toll-like recept. J Immunol 2000;164:554–557.

125 Lehner MD, Morath S, Michelsen KS, Schumann RR, Hartung T: Induction of cross-tolerance by lipopolysaccharide and highly purified lipoteichoic acid via different toll-like receptors independent of paracrine mediators. J Immunol 2001;166:5161–5167.

126 Means TK, Lien E, Yoshimura A, Wang S, Golenbock DT, Fenton MJ: The CD14 ligands lipoarabinomannan and lipopolysaccharide differ in their requirement for Toll-like receptors. J Immunol 1999;163:6748–6755.

127 Hashimoto M, Imamura Y, Morichika T, Arimoto K, Takeuchi O, Takeda K, Akira S, Aoyama K, Tamura T, Kotani S, Suda Y, Kusumoto S: Cytokine-inducing macromolecular glycolipids from *Enterococcus hirae*: Improved method for separation and analysis of its effects on cellular activation. Biochem Biophys Res Commun 2000;273:164–169.

128 Schroder NW, Opitz B, Lamping N, Michelsen KS, Zahringer U, Gobel UB, Schumann RR: Involvement of lipopolysaccharide binding protein, CD14, and Toll-like receptors in the initiation of innate immune responses by *Treponema* glycolipids. J Immunol 2000;165: 2683–2693.

129 Tsuji S, Matsumoto M, Takeuchi O, Akira S, Azuma I, Hayashi A, Toyoshima K, Seya T: Maturation of human dendritic cells by cell wall skeleton of *Mycobacterium bovis* bacillus Calmette-Guerin: Involvement of Toll-like receptors. Infect Immun 2000;68:6883–6890.

130 Ingalls RR, Lien E, Golenbock DT: Membrane-associated proteins of a lipopolysaccharide-deficient mutant of *Neisseria meningitidis* activate the inflammatory response through toll-like receptor 2. Infect Immun 2001;69:2230–2236.

131 Pridmore AC, Wyllie DH, Abdillahi F, Steeghs L, van Der LP, Dower SK, Read RC: A lipopolysaccharide-deficient mutant of *Neisseria meningitidis* elicits attenuated cytokine release by human macrophages and signals via Toll-like receptor (TLR) 2 but not via TLR4/MD2. J Infect Dis 2001;183:89–96.

132 Hajjar AM, O'Mahony DS, Ozinsky A, Underhill DM, Aderem A, Klebanoff SJ, Wilson CB: Cutting edge: Functional interactions between toll-like receptor (TLR) 2 and TLR1 or TLR6 in response to phenol-soluble modulin. J Immunol 2001;166:15–19.

133 Sugawara S, Yang S, Iki K, Hatakeyama J, Tamai R, Takeuchi O, Akashi S, Espevik T, Akira S, Takada H: Monocytic cell activation by nonendotoxic glycoprotein from *Prevotella intermedia* ATCC 25611 is mediated by Toll-like receptor 2. Infect Immun 2001;69:4951–4957.

134 Campos MA, Almeida IC, Takeuchi O, Akira S, Valente EP, Procopio DO, Travassos LR, Smith JA, Golenbock DT, Gazzinelli RT: Activation of Toll-like receptor-2 by glycosylphosphatidylinositol anchors from a protozoan parasite. J Immunol 2001;167:416–423.

135 Jones BW, Means TK, Heldwein KA, Keen MA, Hill PJ, Belisle JT, Fenton MJ: Different Toll-like receptor agonists induce distinct macrophage responses. J Leukoc Biol 2001;69: 1036–1044.

136 Vabulas RM, Ahmad-Nejad P, da Costa C, Miethke T, Kirschning CJ, Hacker H, Wagner H: Endocytosed HSP60s use toll-like receptor 2 (TLR2) and TLR4 to activate the toll/interleukin-1 receptor signaling pathway in innate immune cells. J Biol Chem 2001;276:31332–31339.

137 Li M, Carpio DF, Zheng Y, Bruzzo P, Singh V, Ouaaz F, Medzhitov RM, Beg AA: An essential role of the NF-kappa B/Toll-like receptor pathway in induction of inflammatory and tissue-repair gene expression by necrotic cells. J Immunol 2001;166:7128–7135.

138 Bieback K, Lien E, Klagge IM, Avota E, Schneider-Schaulies J, Duprex WP, Wagner H, Kirschning CJ, Ter M, Schneider-Schaulies S: Hemagglutinin protein of wild-type measles virus activates toll-like receptor 2 signaling. J Virol 2002;76:8729–8736.

139 Ogawa T, Asai Y, Hashimoto M, Uchida H: Bacterial fimbriae activate human peripheral blood monocytes utilizing TLR2, CD14 and CD11a/CD18 as cellular receptors. Eur J Immunol 2002; 32:2543–2550.

140 Flo TH, Ryan L, Latz E, Takeuchi O, Monks BG, Lien E, Halaas O, Akira S, Skjak-Braek G, Golenbock DT, Espevik T: Involvement of toll-like receptor (TLR) 2 and TLR4 in cell activation by mannuronic acid polymers. J Biol Chem 2002;277:35489–35495.

141 Henneke P, Takeuchi O, van Strijp JA, Guttormsen HK, Smith JA, Schromm AB, Espevik TA, Akira S, Nizet V, Kasper DL, Golenbock DT: Novel engagement of CD14 and multiple toll-like receptors by group B streptococci. J Immunol 2001;167:7069–7076.

142 Mambula SS, Sau K, Henneke P, Golenbock DT, Levitz SM: Toll-like receptor (TLR) signaling in response to *Aspergillus fumigatus*. J Biol Chem 2002;277:39320–39326.

143 Sing A, Rost D, Tvardovskaia N, Roggenkamp A, Wiedemann A, Kirschning CJ, Aepfelbacher M, Heesemann J: Yersinia V-antigen exploits toll-like receptor 2 and CD14 for interleukin 10-mediated immunosuppression. J Exp Med 2002;196:1017–1024.

144 Poltorak A, He X, Smirnova I, Liu MY, Huffel CV, Du X, Birdwell D, Alejos E, Silva M, Galanos C, Freudenberg M, Ricciardi-Castagnoli P, Layton B, Beutler B: Defective LPS signaling in C3H/HeJ and C57BL/10ScCr mice: Mutations in Tlr4 gene. Science 1998;282:2085–2088.

145 Means TK, Jones BW, Schromm AB, Shurtleff BA, Smith JA, Keane J, Golenbock DT, Vogel SN, Fenton MJ: Differential effects of a Toll-like receptor antagonist on *Mycobacterium tuberculosis*-induced macrophage responses. J Immunol 2001;166:4074–4082.

146 Qureshi ST, Larivi, Leveque G, Clermont S, Moore KJ, Gros P, Malo D: Endotoxin-tolerant mice have mutations in Toll-like receptor 4 (Tlr4). J Exp Med 1999;189:615–625.

147 Lien E, Chow JC, Hawkins LD, McGuinness PD, Miyake K, Espevik T, Gusovsky F, Golenbock DT: A novel synthetic acyclic lipid A-like agonist activates cells via the lipopolysaccharide/Toll-like receptor 4 signaling pathway. J Biol Chem 2001;276:1873–1880.

148 Kawasaki K, Akashi S, Shimazu R, Yoshida T, Miyake K, Nishijima M: Involvement of TLR4/MD-2 complex in species-specific lipopolysaccharide-mimetic signal transduction by Taxol. J Endotoxin Res 2001;7:232–236.

149 Perera PY, Mayadas TN, Takeuchi O, Akira S, Zaks-Zilberman M, Goyert SM, Vogel SN: CD11b/CD18 acts in concert with CD14 and toll-like receptor (TLR) 4 to elicit full lipopolysaccharide and taxol-inducible gene expression. J Immunol 2001;166:574–581.

150 Haynes LM, Moore DD, Kurt-Jones EA, Finberg RW, Anderson LJ, Tripp RA: Involvement of toll-like receptor 4 in innate immunity to respiratory syncytial virus. J Virol 2001;75: 10730–10737.

151 Ohashi K, Burkart V, Flohe S, Kolb H: Cutting edge: Heat shock protein 60 is a putative endogenous ligand of the toll-like receptor-4 complex. J Immunol 2000;164:558–561.

152 Dybdahl B, Wahba A, Lien E, Flo TH, Waage A, Qureshi N, Sellevold OF, Espevik T, Sundan A: Inflammatory response after open heart surgery: Release of heat-shock protein 70 and signaling through toll-like receptor-4. Circulation 2002;105:685–690.

153 Okamura Y, Watari M, Jerud ES, Young DW, Ishizaka ST, Rose J, Chow JC, Strauss JF: The extra domain A of fibronectin activates toll-like receptor 4. J Biol Chem 2001;276:10229–10233.

154 Smiley ST, King JA, Hancock WW: Fibrinogen stimulates macrophage chemokine secretion through toll-like receptor 4. J Immunol 2001;167:2887–2894.

155 Lee JY, Sohn KH, Rhee SH, Hwang D: Saturated fatty acids, but not unsaturated fatty acids, induce the expression of cyclooxygenase-2 mediated through Toll-like receptor 4. J Biol Chem 2001;276:16683–16689.

156 Shoham S, Huang C, Chen JM, Golenbock DT, Levitz SM: Toll-like receptor 4 mediates intracellular signaling without TNF-alpha release in response to *Cryptococcus neoformans* polysaccharide capsule. J Immunol 2001;166:4620–4626.

157 Wang JE, Warris A, Ellingsen EA, Jorgensen PF, Flo TH, Espevik T, Solberg R, Verweij PE, Aasen AO: Involvement of CD14 and toll-like receptors in activation of human monocytes by *Aspergillus fumigatus hyphae*. Infect Immun 2001;69:2402–2406.

158 Miller YI, Viriyakosol S, Binder CJ, Feramisco JR, Kirkland TN, Witztum JL: Minimally modified LDL binds to CD14, induces macrophage spreading via TLR4/MD-2 and inhibits phagocytosis of apoptotic cells. J Biol Chem 2003:278:1561–1568.

159 Guillot L, Balloy V, McCormack FX, Golenbock DT, Chignard M, Si-Tahar M: Cutting edge: The immunostimulatory activity of the lung surfactant protein-A involves Toll-like receptor 4. J Immunol 2002;168:5989–5992.

160 Biragyn A, Ruffini PA, Leifer CA, Klyushnenkova E, Shakhov A, Chertov O, Shirakawa AK, Farber JM, Segal DM, Oppenheim JJ, Kwak LW: Toll-like receptor 4-dependent activation of dendritic cells by beta-defensin 2. Science 2002;298:1025–1029.
161 Johnson GB, Brunn GJ, Kodaira Y, Platt JL: Receptor-mediated monitoring of tissue well-being via detection of soluble heparan sulfate by Toll-like receptor 4. J Immunol 2002;168:5233–5239.
162 Rassa JC, Meyers JL, Zhang Y, Kudaravalli R, Ross SR: Murine retroviruses activate B cells via interaction with toll-like receptor 4. Proc Natl Acad Sci USA 2002;99:2281–2286.
163 Bulut Y, Faure E, Thomas L, Equils O, Arditi M: Cooperation of Toll-like receptor 2 and 6 for cellular activation by soluble tuberculosis factor and *Borrelia burgdorferi* outer surface protein A lipoprotein: role of Toll-interacting protein and IL-1 receptor signaling molecules in Toll-like receptor 2 signaling. J Immunol 2001;167:987–994.
164 Takeshita F, Leifer CA, Gursel I, Ishii KJ, Takeshita S, Gursel M, Klinman DM: Cutting edge: Role of Toll-like receptor 9 in CpG DNA-induced activation of human cells. J Immunol 2001;167:3555–3558.

Holger Heine, PhD
Div. of Innate Immunity, Department of Immunology and Cell Biology
Research Center Borstel, Parkallee 22, DE–23845 Borstel (Germany)
Tel. +49 4537 188 420, Fax +49 4537 188 697, E-Mail hheine@fz-borstel.de

Kabelitz D, Schröder J-M (eds): Mechanisms of Epithelial Defense.
Chem Immunol Allergy. Basel, Karger, 2005, vol 86, pp 120–135

TLR Signalling and the Function of Dendritic Cells

Hiroaki Hemmi[a], *Shizuo Akira*[a,b]

[a]Department of Host Defense, Research Institute for Microbial Diseases, Osaka University, and [b]ERATO, Japan Science and Technology Corporation, Osaka, Japan

Abstract

The recognition of microbes by innate immune cells initiates activation of the whole immune system. Toll-like receptors (TLRs) are known to recognize various components of invading pathogens. At present, the natural ligands for almost all TLR members have been identified. TLRs are expressed on many types of cells including macrophages and dendritic cells (DCs). The recognition of invading microbes by TLRs on DCs induces proinflammatory cytokine production and enhanced antigen presentation to naive T cells, and finally activates antigen-specific adaptive immune responses. The sequential activation of innate and subsequent adaptive immunity are crucial steps to eradicate invading pathogens. Recently, the TLR signalling pathway has been intensively investigated. Accumulating evidence indicates that, at least, four adaptor molecules are involved in TLR signalling and provide their signalling specificities. Distinct TLR ligands provide distinct activation status and cytokine production patterns for antigen presenting cells, resulting in the induction of differential immune responses. Thus, TLRs are critical molecules to induce not only inflammatory responses but also fine-tuned adaptive immune responses depending on invading pathogens.

Introduction

The mammalian immune system consists of two major types of immunity, innate and adaptive immunity. Adaptive immunity is mediated by B and T lymphocytes recognizing pathogens with high affinity receptors. These receptors are generated through gene rearrangements, which allow generating a random and highly diverse repertoire of receptors. T and B cells undergo clonal selection and expansion with relevant specificities. However, the establishment of adaptive immunity is not rapid enough to eradicate invading microbes. On the other hand, innate immunity plays important roles in the host defence during the early stage of infection. The innate immune system specifically recognizes structural patterns of microbial components, which are conserved among a wide variety of microorganisms [1]. These structural patterns are termed pathogen-associated molecular patterns (PAMPs). PAMPs are pathogen specific and do not present on the host cells, therefore the host innate immune system can discriminate invading pathogens from the self. The recognition of PAMPs in the innate immunity relies on a limited numbers of germline-encoded receptors, called pattern recognition receptors (PRRs). As one of the PPRs, the TLR family is well studied and characterized [2].

What Are Toll-Like Receptors?

The TLR family has been phylogenetically conserved from insects to mammals [1, 2]. TLRs are characterized as a type I transmembrane protein with leucine-rich repeats in the extracellular domain and a cytoplasmic Toll/interleukin (IL)-1 receptor homology (TIR) domain. So far, ten TLR members have been identified in humans. Accumulating evidences indicate that TLRs can recognize various microbial components (table 1). The recognition of PAMPs by TLRs induces secretion of inflammatory cytokines such as TNF-α, IL-6, and IL-1β by various types of cells including macrophages (Mϕs) and dendritic cells (DCs). In addition, stimulation with TLR ligands also induces the functional maturation of DCs, especially an enhancement of their antigen presenting ability to naive T cells, resulting in effective activation of acquired immunity.

Toll-Like Receptors Recognize Various Molecules

Many reports indicate that TLR family can recognize various components of microbes; a selection of microbial components that activate TLRs are listed in table 1. The ligand for TLR4 was first identified. It has been reported that LPS-hyporesponsive mouse strain, C3H/HeJ and C57BL/10ScCr, have a point mutation in TIR domain of TLR4 gene and are null mutation of the gene, respectively. Furthermore, TLR4 gene knockout mice did not show any response to LPS. At present, it has been shown that various bacterial components, such as lipoteichoic acid from gram negative bacteria and HSP60 from *Chlamydia*, can activate immune cells through TLR4. TLR2 also mediates immune responses to cell-wall components from many pathogens, including lipoproteins, lipoarabinomannan from mycobacteria, glycosylphosphatidylinositol anchors from *Trypanosoma cruzi*, and zymosan from fungi and glycolipids from *Treponema maltophilum*. Interestingly, TLR2 can recognize lipoproteins by cooperation with TLR1 or

Table 1. The TLR family members recognize various molecules

TLRs	Recognized molecules and microbes
TLR1/TLR2/TLR6	peptidoglycan (gram-positive bacteria) phenol-soluble modulin (*Neisseria meningitis*) [TLR1/TLR2] 19 kDa lipoprotein (mycobacteria) [TLR1/TLR2] lipopeptides (mycoplasma) [TLR2/TLR6] zymosan (yeast) GPI anchors (*Trypanosoma cruzi*)
TLR3	double-stranded RNA (virus)
TLR4	LPS (gram-negative bacteria) lipoteichoic acid (gram-positive bacteria) F protein (respiratory syncytial virus) Taxol (Plants) heat shock protein 60 (Host)
TLR7/TLR8	imidazoquinolines (synthetic compounds) single-stranded RNA influenza virus, vascular stomatitis virus (mouse TLR7)
TLR9	CpG DNA(virus, bacteria) herpes simplex virus

TLR2 can recognize various bacterial components together with TLR1 or TLR6. The combination is indicated in square brackets.

TLR6. For example, cytokine production in response to mycoplasmal lipopeptides is abolished in TLR2- or TLR6-deficient cells but retained in TLR1-deficient cells. On the other hand, bacterial lipopeptides can activate TLR6-deficient cells but not TLR1- or TLR2-deficient cells. Mycoplasmal and bacterial lipopeptides are structurally different; mycoplasmal lipopeptides have diacylated cysteine residues at their amino-terminus, while the corresponding cysteine residue in bacterial lipopeptides is triacylated. Thus, TLR2 together with TLR1 and TLR6 can discriminate these structural differences. TLR1, TLR2, and TLR6 are located in close proximity to each other on the same chromosome. TLR5 can recognize flagellin, one of the structural proteins of bacterial flagellae.

TLR7 and TLR8 genes are located on the X chromosome and show high homology with each other. It has been reported that synthetic anti-viral compounds, imidazoquinolines such as R-848 and imiquimod, can induce immune cell activation via TLR7 in mice. In humans, both TLR7 and TLR8 recognize R-848. In addition to the imidazoquinolines, some guanine nucleotide analogues also activate TLR7. Moreover, TLR7 recognizes single-stranded RNAs (ssRNA) or ssRNA viruses, including uridine-rich or uridine/guanosine-rich oligonucleotides, influenza virus, and vascular stomatitis virus [3–5]. In addition, TLR7 can also recognize usual mRNAs such as mouse splenic RNA and in vitro transcribed mRNA, suggesting that TLR7 cannot discriminate structures of ssRNAs. Since TLR7 and 8 seem to be located within the endosomes, the localization of ssRNA in intracellular compartments might be important for TLR7-mediated cell activation.

TLR9 shows high homology with TLR7 and TLR8 as compared with the other members. TLR9 mediates immune responses to immunostimulatory DNA containing unmethylated CpG motifs, including bacterial genomic DNA and synthetic oligodeoxynucleotides (CpG DNAs). CpG DNAs have strong immunostimulatory activity such as induction of inflammatory cytokine production and T helper (Th) 1-polarizing immune responses. The presence of unmethylated CpG motif in CpG DNAs is responsible for their immunostimulatory activity. The CpG motif is abundant in bacterial genomes, while the frequency of the motif is suppressed and highly methylated in mammalian DNA. Therefore, mammalian DNA does not show the same immunostimulatory activity as that of bacterial DNA. TLR9-deficient mice show defective responses to CpG DNA and expression of the TLR9 gene confers the responsiveness to CpG DNA in vitro. In addition to bacterial and synthetic DNA, TLR9 also mediates herpes simplex virus induced IFN-α production. TLR9 seems to be located in endosomes and endosomal maturation is prerequisite to initiate its signalling.

Thus, TLRs can recognize various molecules, and the recognition leads to production of proinflammatory cytokines in DCs and Mφs. In addition, to proinflammatory cytokine production, some TLR ligands stimulate the cells to secretion of type I IFNs. Interestingly, TLRs can be classified into three groups in terms of type I IFN production. TLR2 ligands cannot induce IFN-β production in DCs and M s. TLR4 ligands can induce IFN-β, but not IFN-α, production, whereas TLR3 ligand, poly(I:C), can stimulate production of both type I IFNs in DCs. Activation of TLR7 or TLR9 stimulates IFN-β production. In addition to IFN-β, TLR7 and TLR9 ligands also induce IFN-α production in a subset of DCs, plasmacytoid DCs. These facts suggest that the TLR signalling pathway is diverged, at least, in terms of type I IFN production.

Signalling Pathway of TLRs

MyD88-Dependent Pathway

The signalling event from TLR family members is initiated from TIR domain. The activation of TLRs induces recruitment of TIR domain containing adaptor proteins (Fig. 1). MyD88 is a common adaptor protein containing TIR domain and associated with TIR domain of TLR and IL-1 receptor family.

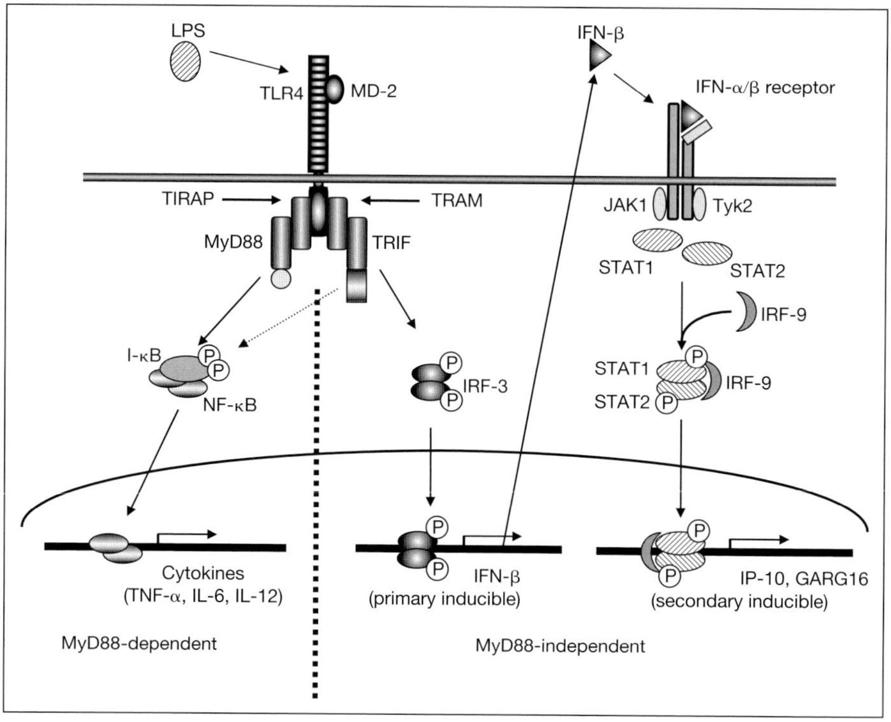

Fig. 1. MyD88-dependent and -independent signalling pathway in TLR4 signalling. A MyD88-dependent pathway leads to production of inflammatory cytokines and rapid activation of NF-κB and MAP kinases. The analysis of TRIF mutant cells shows that the activation of a MyD88-dependent pathway is not sufficient for proinflammatory cytokine production in TLR4 signalling. A MyD88-independent pathway induces slow activation of NF-κB and MAP kinases as well as IRF-3. The activation of IRF-3 induces production of IFN-β. Secreted IFN-β stimulates IFN-α/β receptor which leads to activation of STAT1 and STAT2, resulting in induction of IFN-inducible genes such as IP-10 and GARG-16.

The ligation of cognate ligands to TLRs induces the recruitment of MyD88 and sequential activation of signalling molecules such as IL-1 receptor associated kinase (IRAK) family and TNF receptor associated factor (TRAF) 6. Subsequently, several signalling pathways are activated in parallel, which include activation of IκB kinase complexes and mitogen-activating kinase (MAP kinase) cascade. Finally, transcription factors NF-κB and AP-1 are activated and translocated to the nucleus, where they induce gene expression.

The analysis of MyD88-deficient mice revealed that MyD88 plays a crucial role in TLR/interleukin-1 receptor signalling pathways. Mφs and DCs from MyD88-deficient mice were completely defective in production of inflammatory cytokines such as TNF-α, IL-6, and IL-12 p40 in response to all

TLR ligands. Also, MyD88-deficient splenocytes did not show proliferative responses to stimulation with IL-1, LPS, or CpG DNA. Furthermore, activation of signalling molecules such as NF-κB and MAP kinases was not observed in MyD88-deficient cells in response to TLR2, TLR7, or TLR9 ligands, suggesting that TLR2, TLR7, and TLR9 do not have the MyD88-independent signalling pathway. In sharp contrast, poly(I:C) or LPS can induce activation of signalling molecules in MyD88-deficient cells, suggesting that TLR3 and TLR4 have a MyD88-independent signalling pathway [2, 6].

MyD88-Independent Pathway

In MyD88-deficient cells, production of proinflammatory cytokines such as TNF-α and IL-6 is not observed upon LPS stimulation, whereas activation of MAP kinases and NF-κB is clearly induced [2]. Furthermore, expression of IFN-β and IFN-inducible genes such as IP-10, GARG-16, or IRG-1 are observed in LPS-stimulated MyD88-deficient cells. The expression of these genes is known to be mediated by interferon regulatory factor (IRF)-3 in virus infection. It has been reported that activation of IRF-3 is observed in both LPS-stimulated wild-type and MyD88-deficient cells [2]. Moreover, IRF-3-deficient cells do not produce IFN-β in response to LPS [7]. Thus, IRF-3 plays important role in LPS-stimulated cells. Expression of all IFN-inducible genes is abolished in IFN-α/β receptor deficient cells, suggesting that IFN-β and its signalling are responsible for induction of IFN-inducible genes (fig. 1). In addition to IFN-β production, MyD88-deficient DCs show upregulation of costimulatory molecules on their surface in response to LPS. Similar to the LPS responses, TLR9 ligand can induce IFN-β and IFN-inducible genes and up-regulation of CD40 in DCs. But, these responses are dependent on MyD88. Thus, TLR4 has another signalling pathway independent of MyD88.

TIRAP/Mal

In addition to MyD88, several TIR domain containing adaptor proteins have been identified and shown to be involved in the TLR signalling pathway (Fig. 2). Toll-interleukin 1 receptor (TIR) domain-containing adaptor protein (TIRAP)/MyD88-adaptor-like (Mal) have been identified and characterized [8]. TIRAP is originally shown to be involved in MyD88-independent pathway of TLR4 signalling [8, 9]. In vitro analysis showed that a dominant negative form of TIRAP inhibits TLR4- but not TLR9- or IL-1-induced NF-κB activation and TLR4-induced IRF-3 activation. Moreover, cell membrane permeable TIRAP peptide inhibits cytokine production as well as maturation of DCs upon LPS stimulation, suggesting that TIRAP is implicated in the MyD88-independent pathway. To investigate the physiological function of TIRAP, TIRAP-deficient mice have been generated and characterized [10, 11]. Consistent with in vitro studies, TIRAP

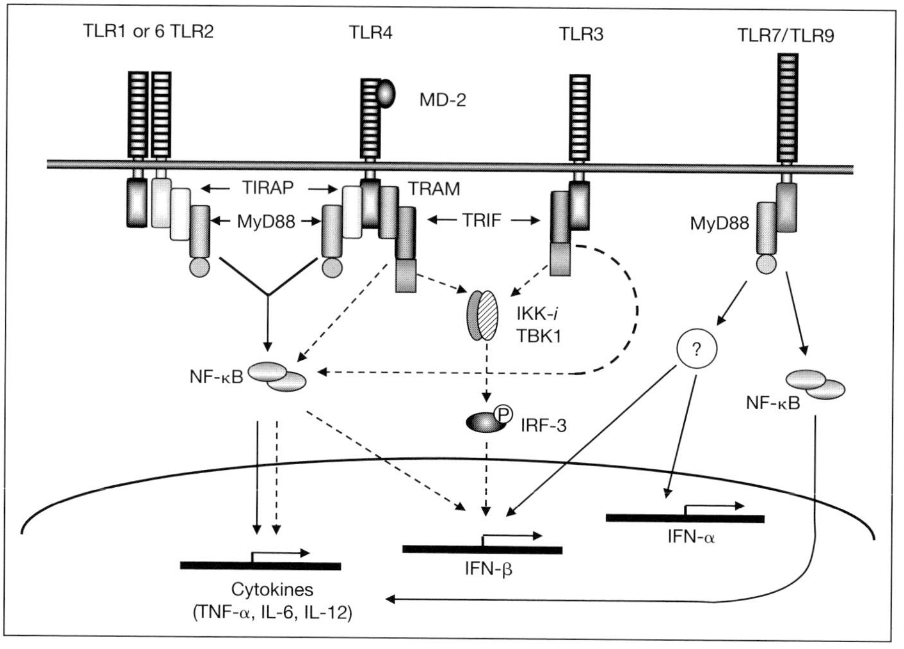

Fig. 2. TLR signalling pathways. Differential adaptor molecules provide specific signalling pathway in each TLR family members. TLR2 cannot induce IFN-β production, whereas TLR3 and TLR4 induce IFN-β through adaptor protein TIRF- and IKK-related kinases. The activation of TLR7 and TLR9 induce IFN-β as well as IFN-α production in particular DC subsets. TLR7- and TLR9-induced cytokine production and activation of signalling molecules are dependent on MyD88. The molecules involved in IFN-α/β production remain unclear in TLR7/9 signalling.

deficiency resulted in impaired responses such as proliferation of splenocytes and inflammatory cytokine production by Mφs to LPS stimulation. Unexpectedly, TIRAP-deficient cells did not respond to the TLR2 ligand, peptidoglycan and mycoplasmal lipopeptides. In contrast to the defective responses induced by TLR4 or TLR2 ligands in TIRAP-deficient cells, the responses to TLR3, TLR5, TLR7, and TLR9 ligands were comparable to those observed in wild-type cells. Also, the responses to IL-1 and IL-18 were not impaired in TIRAP-deficient cells. Thus, TIRAP plays important roles in the MyD88-dependent pathway shared by TLR2 and TLR4 signalling. Despite these defective responses induced by the MyD88-dependent pathway in TIRAP-deficient cells, IRF-3 activation and expression of IFN-β and IFN-inducible genes were clearly observed. Furthermore, IFN-β and IFN-inducible genes could be induced in MyD88/TIRAP doubly deficient cells in response to LPS [10], demonstrating that TIRAP is not involved in the MyD88-independent pathway downstream of TLR4.

TRIF

Through database search screening, some novel proteins harboring the TIR domain with unknown functions were identified. One of these molecules was termed TIR domain-containing adaptor-inducing IFN-β (TRIF) [12]. TRIF was also named TIR-containing adaptor molecule (TICAM)-1 and identified through the yeast two-hybrid screening using intracellular domain of TLR3 as bait [13]. Human TRIF consists of 712 amino acid and is larger than other TIR containing adaptor molecules such as MyD88 (296 amino acids) and TIRAP (232 amino acids). Overexpression of TRIF in human embryonic kidney 293 cells resulted in the activation of NF-κB, albeit at a low level compared with MyD88- or TIRAP-induced activation. A reporter assay to monitor IFN-β promoter activation revealed that overexpression of TRIF induced activation of the IFN-β promoter, whereas neither MyD88 nor TIRAP overexpression activated the IFN-β promoter. Furthermore, TRIF could directly associate with TLR3, and a dominant negative form of TRIF, but not MyD88 or TIRAP, inhibited TLR3-induced activation of the IFN-β promoter and NF-κB, suggesting that TIRF is involved in the TLR3-mediated MyD88-independent pathway.

In addition to these in vitro studies, mice carrying the mutated *Trif* gene have been generated and analyzed independently by two different groups [14, 15]. One group has generated TRIF-deficient mice by conventional gene targeting, and the other group has identified germline mutant mice termed *LPS2* mice using N-ethyl-N-nitrosourea (ENU)-mediated mutagenesis. Both mutant mice showed similar phenotypes. TRIF mutant cells did not induce IRF-3 activation or the expression of IFN-β and IFN-inducible genes in response to poly(I:C) or LPS. In addition, LPS-induced cytokine production was also impaired. LPS-induced NF-κB and MAP kinase activation was observed in TRIF-deficient cells to the same extent as that induced in wild-type cells. In contrast, poly(I:C)-induced NF-κB activation was severely impaired. LPS-induced NF-κB and MAP kinase activation in TRIF-deficient cells is mediated by MyD88, since the activation was not observed at all in MyD88/TRIF doubly deficient cells. Thus, TRIF specifically mediates the MyD88-independent pathway downstream of TLR4 and TLR3 (Fig. 2).

Interestingly, TRIF mutant mice were also defective in proinflammatory cytokine production upon LPS stimulation. TRIF mutant cells could produce proinflammatory cytokines such as TNF-α, IL-6, and IL-12 p40 in response to TLR2, TLR7, or TLR9 ligands. On the other hand, these proinflammatory cytokine productions were severely impaired in response to LPS. Thus, activation of either MyD88-dependent or TRIF-dependent pathway is not sufficient for inflammatory cytokine productions upon LPS stimulation.

The transcription factor IRF-3 plays a crucial role in the MyD88-independent pathway. It has been also reported that two IκB kinase (IKK)-related kinases are involved in the activation of IRF-3 in viral infection as well as in the MyD88-independent signalling pathway (fig. 2). One kinase is called inducible IKK (IKK-*i*) and the other is TANK-binding kinase-1 (TBK1). IKK-*i*, also called IKK-ε, was originally identified as an LPS-inducible kinase. TBK1 is also known as NF-κB-activating kinase (NAK) or TRAF2-associated kinase (T2K) [16]. These two kinases share homology with IKK-α and IKK-β. Both kinases have been reported to interact with TRAF2 and the TRAF-binding protein TANK/I-TRAF, and function upstream of IKK-α and IKK-β. Deficiency of TBK1 in mice resulted in the embryonic lethality caused by massive liver degeneration, and TBK1-deficient embryonic fibroblasts (EFs) showed reduced expression of certain genes regulated by NF-κB. Thus, these two IKK-related kinases have been implicated in NF-κB activation. Recently, two groups have shown that IKK-*i* and TBK1 phosphorylate and activate IRF-3 and IRF-7, leading to the transcriptional activation of genes for IFN-β, RANTES, and ISG54 in viral infection as well as in a TRIF-dependent signalling pathway [17, 18]. It has also been reported that TLR4-induced signalling events leading to activation of NF-κB and IRF-3 are diverged at TRIF, and depend on TRAF6 and TBK1, respectively [19]. The analysis of TBK1- or IKK-*i*-deficient cells focused on IFN-β induction has been also reported [20]. TBK1-deficient embryonic fibroblasts (EFs) showed marked decrease in the induction of the IFN-β and IFN-inducible genes in response to LPS and virus infection. In addition, TBK1-deficient cells showed delayed induction of IFN-β and IFN-inducible genes in response to poly(I:C). The activation of IRF-3 was also diminished. In this report, poly(I:C) was complexed with cationic lipids and then added to the cultures. The poly(I:C)-cationic lipid complexes induced TLR3-independent expression of IFN-β and IFN-inducible genes, suggesting that TBK1 is involved in TRIF-dependent as well as -independent signalling pathway. In contrast to impairment of IFN-β induction in TBK1-deficient cells, IKK-*i*-deficient cells did not show impaired responses to LPS, poly(I:C), or viral infection. However, in TBK1/IKK-*i* doubly deficient cells, the delayed IFN-β induction and diminished IRF-3 activation observed in TBK1 single deficient cells were completely abolished, suggesting that IKK-*i* is also involved in poly(I:C) signalling which induces IRF-3 activation and IFN-β induction. Similarly, other groups have reported that TBK1 is involved in IRF-3 activation and induction of IFN-β in a TRIF-dependent signalling (TLR3 and TLR4) and viral infection [20–22].

TRAM

TRIF-related adaptor molecule (TRAM, also called TICAM-2 or TIRP) has also been identified through a database search focusing on the TIR domain [23–26]. TRAM consists of 235 amino acids. Initial experiments suggested that TRAM/TIRP is involved in IL-1 receptor signalling to activate NF-κB [26]. In addition to NF-κB activation, other groups reported overexpression of TRAM-induced activation of the IFN-β promoter, and that TRAM can bind to TLR4 [23–25]. Fitzgerald et al. [24] reported that the dominant negative form of TRIF inhibited TRAM-induced NF-κB activation and the dominant negative form of TRAM could not inhibit TRIF-induced NF-κB activation, indicating that TRAM functions upstream of TIRF in TLR4 signalling. The analysis of TRAM-deficient mice showed that TRAM is specifically implicated in the MyD88-independent pathway downstream of TLR4 signalling. The responses to TLR2, TLR7, and TLR9 were indistinguishable between wild-type and TRAM-deficient cells. In terms of TLR4-mediated stimulation, TRAM-deficient cells showed the same phenotypes as that observed in TRIF-deficient cells, i.e. impaired IRF-3 activation and IFN-β induction. Interestingly, poly(I:C)-induced activation of IRF-3 and expression of IFN-β and IFN-inducible genes were comparable to that observed in wild-type cells, suggesting that TRAM specifically mediates the MyD88-independent pathway of TLR4 signalling (Fig. 2).

The Role of TLR Family in the Host Defence

TLRs Stimulate DCs to Induce T Cell Activation

TLRs can play important roles not only in sensing pathogens and inducing inflammatory responses but also in linking innate and adaptive immunity (Fig. 3). TLRs are also expressed on DCs, known as professional antigen-presenting cells, and the activation of TLRs induces an enhancement of their antigen-presenting activity. DCs in the periphery capture foreign antigens and migrate into the T cell areas in the lymphoid tissues, where they process and present antigen to naive T cells. This can initiate the activation of antigen-specific immunity. Antigen presentation alone can stimulate antigen-specific T cell clones, but it is not sufficient to induce their clonal expansion. Additional signals from costimulatory molecules such as CD80/CD86 are indispensable for their clonal expansion. T cells that receive TCR signalling in the absence of costimulatory signals become anergic, whereas the cells that receive signals from TCR together with costimulatory molecules are activated and expanded.

In addition to T cell expansion, T cell differentiation is also important to eradicate invading microbes. CD4+ naive T cells can differentiate into two distinct types of helper T cells, type I and type II (hereafter Th1 and Th2, respectively). Th1 cells mainly secrete the effector cytokine IFN-γ and are involved in cellular immunity. Th2 cells produce IL-4, IL-5, IL-10, and IL-13 and are involved in humoral immunity. Differentiation of Th cells can be controlled by

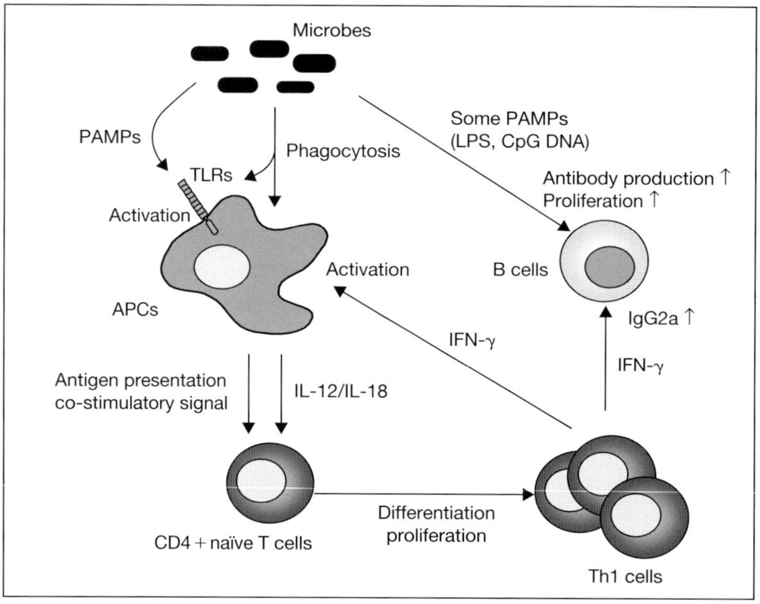

Fig. 3. Immune cell activation via TLR family. Invading pathogens are captured by antigen presenting cells (APCs) such as DCs and MIs, which process and present antigens on MHC class II. The recognition of PAMPs by TLRs induces DC maturation including upregulation of costimulatory molecules, and production of inflammatory and Th1 inducing cytokines. These maturated DCs result in differentiation and expansion of antigen-specific Th1 cells. Th1 cells produce effector cytokine IFN-γ, which induces the activation of DCs and M s and an immunoglobulin class switch to IgG2a in B cells. Furthermore, some PAMPs can directly stimulate B cells to induce IL-6 and immunoglobulin production, and proliferation.

DCs, depending on the particular DC subset, DC maturation stage, the DC to T cell ratio, or antigen doses [27, 28]. Bacterial infection activates DCs via stimulation of some TLR family members and mainly induces production of Th1 inducing cytokines such as IL-12, resulting in induction of Th1 immune response. Especially TLR7 and TLR9 ligands are known to induce strong Th1 responses. On the other hand, it has been reported that the TLR2 ligand, Pam3Cys, and the TLR5 ligand, flagellin, can induce Th2 cells in vitro [29, 30]. It has also been reported that treatment with high-dose LPS induces Th1 cells, while low-dose LPS stimulates Th2 responses [29]. Furthermore, transfer of dsRNA-stimulated DCs induces production of all subtypes of IgG including IgG1 and IgG2a, suggesting that poly(I:C) enhance both Th1 and Th2 immune responses [31]. Thus, TLR ligands affect Th1 and Th2 responses through DCs.

LPS-Stimulated MyD88-Deficient DCs

As discussed above, TLR4 signalling is also transduced in a MyD88-independent fashion. MyD88-deficient DCs show up-regulation of co-stimulatory molecules and secretion of IFN-β upon LPS stimulation, while they do not produce cytokines such as TNF-α and IL-6, and IL-12 [32]. These responses are completely abolished in CpG DNA-stimulated MyD88-deficient DCs. Interestingly, LPS-stimulated MyD88-deficient DCs could induce Th2 cells when co-cultured with naive CD4+ T cells under the experimental conditions which LPS-stimulated wild-type cells induce Th1 cells. Because LPS-stimulated IL-12- or IL-18-deficient or IL-12/IL-18 doubly deficient cells did not induce Th1 cells as well as Th2 cells, the Th2 cell-inducing ability of MyD88-deficient DCs is not simply due to the lack of these Th1-inducing cytokine productions. Recently, it has been reported that Notch ligand Jagged-1, but not Delta, can induce Th2 cells and Jagged-1 is induced in LPS-stimulated MyD88-deficient DCs [33, 34]. Thus, two major biological effects induced by LPS, cytokine production and antigen presentation differ in their requirement for MyD88.

DC Subset-Dependent Cytokine Production

DCs are highly heterogeneous and can be grouped into several subsets according to the expression of their surface markers and their biological functions. Interestingly, the expression pattern of TLRs differs between DC subsets. In humans, one DC subset termed plasmacytoid DCs (PDCs), which contain IFN-α producing cells in viral infection, expresses TLR7 and TLR9 but not TLR1–6 or TLR8. In contrast, the other DC subset, myeloid DCs (MDCs) expresses TLR1–7, but not TLR9. Similar expression patterns are also observed in mouse DCs except for TLR9 expression, i.e. all mouse DCs express TLR9. Furthermore, PDCs and MDCs produce differential cytokines in response to TLR ligands. PDCs can produce IFN-α, but not IL-12, in response to TLR7 or TLR9 ligands, while MDCs secrete IL-12, but not IFN-α, in response to TLR4 or TLR7/8 ligands.

In addition to DC subset-dependent cytokine productions, it has been reported that a subset of DCs shows differential cytokine production depending on the type of CpG DNAs (TLR9 ligands). CpG DNAs can be classified into two or three types according to their sequences, structures, and biological activities. One type of CpG DNAs is phosphorothioate-modified oligodeoxynucleotides called K-type CpG DNAs or CpG-B (conventional CpG DNAs), which induce splenic B cell proliferation, DC maturation, and cytokine production from a variety of immune cells [35]. Through the screening of a variety of CpG DNAs, another type of CpG DNAs with a distinct function was identified [35, 36]. These are termed D-type CpG DNAs or CpG-A (A/D-type CpG DNAs) and are structurally different from conventional CpG DNAs, since they carry a phosphorothioate-modified poly-guanosine (polyG) stretch at the 5′ and 3′-ends

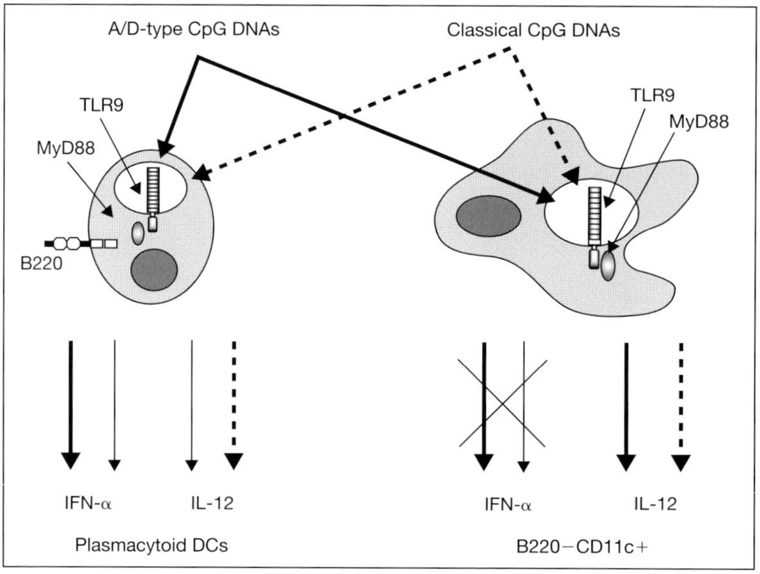

Fig. 4. Mouse DC subset-dependent cytokine production in response to TLR9 ligands. Both types of CpG DNAs are taken up by DCs and stimulate the cell through TLR9. Classical CpG DNAs induce IL-12 production from both subsets of mouse DCs (dotted line). On the other hand, A/D-type DNA stimulates plasmacytoid DCs to produce IFN-α, and B220−CD11c+ DCs to produce IL-12, but not IFN-α, respectively. These cytokine productions are dependent on TLR9 and MyD88.

and a phosphodiester-backbone CpG motif at the central position. The function of A/D-type CpG DNAs has been extensively characterized in the human system. A/D-type CpG DNAs have higher activity to induce IFN-α production from plasmacytoid DC (PDC) and IFN-γ from natural killer cells.

Mouse DCs can be divided into two groups according to the expression of B220, B220+ and B220− DCs. B220+ DCs contain the counterpart of human PDCs. Both DC subsets express TLR9 and respond to conventional CpG DNAs, which can induce up-regulation of costimulatory molecules and IL-12 p40 production. On the other hand, upon A/D-type CpG DNA stimulation, both DC subsets showed upregulation of costimulatory molecules, but each subset produced different cytokines from one another; B220+ DCs produced mainly IFN-α and B220− DCs secreted IL-12 p40 but not IFN-α [38]. Furthermore, all of these responses were completely abolished in MyD88- or TLR9-deficient cells. Thus, PDCs and B220− DCs secrete differential cytokines in response to the same TLR9 ligands (A/D-type CpG DNAs), and different TLR9 ligands induce distinct cytokine production by PDCs (Fig. 4). It still remains unknown what mechanisms regulate differential cytokine production in DCs by TLR9 ligands.

Conclusion

The identification of a TIR domain containing adaptor molecules provides new insights into TLR biology as to how signalling specificity among TLR family is regulated. TLR2 needs two adaptors to transduce signals. TLR4 utilizes four adaptors. TLR3 does not require MyD88 for its signalling pathway. TLR3- and TLR4-induced IFN-β production is independent of MyD88. On the other hand, TLR7 and TLR9 signalling is abolished in the absence of MyD88 and therefore, type I IFN production via these signalling pathways depends on MyD88. It still remains unclear how TLR7 and TLR9 regulate type I IFN production. Does another adaptor protein exist which is involved in type I IFN induction which is utilized in TLR9 signalling?

As discussed above, DCs are differentially activated by individual TLRs, and their activation status is dependent on DC subsets and TLR ligands. However, the molecular mechanisms of how Th1-inducing and Th2-inducing abilities of DCs are regulated by TLR signalling remain unclear. Further knowledge of TLRs and DCs will provide us with more effective vaccination strategies with fewer side effects as well as more rational manipulation of immune diseases.

References

1 Janeway CA Jr, Medzhitov R: Innate immune recognition. Annu Rev Immunol 2002;20: 197–216.

2 Takeda K, Kaisho T, Akira S: Toll-like receptors. Annu Rev Immunol 2003;21:335–376.

3 Heil F, Hemmi H, Hochrein H, Ampenberger F, Kirschning C, Akira S, Lipford G, Wagner H, Bauer S: Species-specific recognition of single-stranded RNA via toll-like receptor 7 and 8. Science 2004;303:1526–1529.

4 Diebold SS, Kaisho T, Hemmi H, Akira S, Reis e Sousa C: Innate antiviral responses by means of TLR7-mediated recognition of single-stranded RNA. Science 2004;303:1529–1531.

5 Lund JM, Alexopoulou L, Sato A, Karow M, Adams NC, Gale NW, Iwasaki A, Flavell RA: Recognition of single-stranded RNA viruses by Toll-like receptor 7. Proc Natl Acad Sci USA 2004;101:5598–5603.

6 Alexopoulou L, Holt AC, Medzhitov R, Flavell RA: Recognition of double-stranded RNA and activation of NF-kappaB by Toll-like receptor 3. Nature 2001;413:732–738.

7 Sakaguchi S, Negishi H, Asagiri M, Nakajima C, Mizutani T, Takaoka A, Honda K, Taniguchi T: Essential role of IRF-3 in lipopolysaccharide-induced interferon-beta gene expression and endotoxin shock. Biochem Biophys Res Commun 2003;306:860–866.

8 Horng T, Barton GM, Medzhitov R: TIRAP: an adapter molecule in the Toll signaling pathway. Nat Immunol 2001;2:835–841.

9 Fitzgerald KA, Palsson-McDermott EM, Bowie AG, Jefferies CA, Mansell AS, Brady G, Brint E, Dunne A, Gray P, Harte MT, McMurray D, Smith DE, Sims JE, Bird TA, O'Neill LA: Mal (MyD88-adapter-like) is required for Toll-like receptor-4 signal transduction. Nature 2001;413: 78–83.
10 Yamamoto M, Sato S, Hemmi H, Sanjo H, Uematsu S, Kaisho T, Hoshino K, Takeuchi O, Kobayashi M, Fujita T, Takeda K, Akira S: Essential role for TIRAP in activation of the signalling cascade shared by TLR2 and TLR4. Nature 2002;420:324–329.
11 Horng T, Barton GM, Flavell RA, Medzhitov R: The adaptor molecule TIRAP provides signalling specificity for Toll-like receptors. Nature 2002;420:329–333.
12 Yamamoto M, Sato S, Mori K, Hoshino K, Takeuchi O, Takeda K, Akira S: A novel Toll/IL-1 receptor domain-containing adapter that preferentially activates the IFN-β promoter in the Toll-like receptor signaling. J Immunol 2002;169:6668–6672.
13 Oshiumi H, Matsumoto M, Funami K, Akazawa T, Seya T: TICAM-1, an adaptor molecule that participates in Toll-like receptor 3-mediated interferon-beta induction. Nat Immunol 2003;4:161–167.
14 Yamamoto M, Sato S, Hemmi H, Hoshino K, Kaisho T, Sanjo H, Takeuchi O, Sugiyama M, Okabe M, Takeda K, Akira S: Role of adaptor TRIF in the MyD88-independent toll-like receptor signaling pathway. Science 2003;301:640–643.
15 Hoebe K, Du X, Georgel P, Janssen E, Tabeta K, Kim SO, Goode J, Lin P, Mann N, Mudd S, Crozat K, Sovath S, Han J, Beutler B: Identification of Lps2 as a key transducer of MyD88-independent TIR signalling. Nature 2003;424:743–748.
16 Peters RT, Maniatis T: A new family of IKK-related kinases may function as I kappa B kinase kinases. Biochim Biophys Acta 2001;1471:M57–62.
17 Sharma S, tenOever BR, Grandvaux N, Zhou GP, Lin R, Hiscott J: Triggering the interferon antiviral response through an IKK-related pathway. Science 2003;300:1148–1151.
18 Fitzgerald KA, McWhirter SM, Faia KL, Rowe DC, Latz E, Golenbock DT, Coyle AJ, Liao SM, Maniatis T: IKKepsilon and TBK1 are essential components of the IRF3 signaling pathway. Nat Immunol 2003;4:491–496.
19 Sato S, Sugiyama M, Yamamoto M, Watanabe Y, Kawai T, Takeda K, Akira S: Toll/IL-1 receptor domain-containing adaptor inducing IFN-beta (TRIF) associates with TNF receptor-associated factor 6 and TANK-binding kinase 1, and activates two distinct transcription factors, NF-kappa B and IFN-regulatory factor-3, in the Toll-like receptor signaling. J Immunol 2003;171:4304–4310.
20 McWhirter SM, Fitzgerald KA, Rosains J, Rowe DC, Golenbock DT, Maniatis T: IFN-regulatory factor 3-dependent gene expression is defective in Tbk1-deficient mouse embryonic fibroblasts. Proc Natl Acad Sci USA 2004;101:233–238.
21 Hemmi H, Takeuchi O, Sato S, Yamamoto M, Kaisho T, Sanjo H, Kawai T, Hoshino K, Takeda K, Akira S: The roles of two IkappaB kinase-related kinases in Lipopolysaccharide and double stranded RNA signaling and viral infection. J Exp Med 2004;199:1641–1650.
22 Perry AK, Chow EK, Goodnough JB, Yeh WC, Cheng G: Differential requirement for TANK-binding kinase-1 in type I interferon responses to Toll-like receptor activation and viral infection. J Exp Med 2004;199:1651–1658.
23 Yamamoto M, Sato S, Hemmi H, Uematsu S, Hoshino K, Kaisho T, Takeuchi O, Takeda K, Akira S: TRAM is specifically involved in the Toll-like receptor 4-mediated MyD88-independent signaling pathway. Nat Immunol 2003;4:1144–1150.
24 Fitzgerald KA, Rowe DC, Barnes BJ, Caffrey DR, Visintin A, Latz E, Monks B, Pitha PM, Golenbock DT: LPS-TLR4 signaling to IRF-3/7 and NF-kappaB involves the toll adapters TRAM and TRIF. J Exp Med 2003;198:1043–1055.
25 Oshiumi H, Sasai M, Shida K, Fujita T, Matsumoto M, Seya T: TIR-containing adapter molecule (TICAM)-2, a bridging adapter recruiting to toll-like receptor 4 TICAM-1 that induces interferon-beta. J Biol Chem 2003;278:49751–49762.
26 Bin LH, Xu LG, Shu HB: TIRP, a novel Toll/interleukin-1 receptor (TIR) domain-containing adapter protein involved in TIR signaling. J Biol Chem 2003;278:24526–24532.
27 Moser M, Murphy KM: Dendritic cell regulation of TH1-TH2 development. Nat Immunol 2000; 1:199–205.
28 Pulendran B, Banchereau J, Maraskovsky E, Maliszewski C: Modulating the immune response with dendritic cells and their growth factors. Trends Immunol 2001;22:41–47.

29 Mazzoni A, Segal DM: Controlling the Toll road to dendritic cell polarization. J Leukoc Biol 2004;75:721–730.

30 Didierlaurent A, Ferrero I, Otten LA, Dubois B, Reinhardt M, Carlsen H, Blomhoff R, Akira S, Kraehenbuhl JP, Sirard JC: Flagellin promotes myeloid differentiation factor 88-dependent development of Th2-type response. J Immunol 2004;172:6922–6930.

31 Le Bon A, Schiavoni G, D'Agostino G, Gresser I, Belardelli F, Tough DF: Type I interferons potently enhance humoral immunity and can promote isotype switching by stimulating dendritic cells in vivo. Immunity 2001;14:461–470.

32 Kaisho T, Akira S: Regulation of dendritic cell function through Toll-like receptors. Curr Mol Med 2003;3:373–385.

33 Tanigaki K, Tsuji M, Yamamoto N, Han H, Tsukada J, Inoue H, Kubo M, Honjo T: Regulation of alphabeta/gammadelta T cell lineage commitment and peripheral T cell responses by Notch/RBP-J signaling. Immunity 2004;20:611–622.

34 Amsen D, Blander JM, Lee GR, Tanigaki K, Honjo T, Flavell RA: Instruction of distinct CD4 T helper cell fates by different notch ligands on antigen-presenting cells. Cell 2004;117:515–526.

35 Krieg AM: CpG motifs in bacterial DNA and their immune effects. Annu Rev Immunol 2002;20:709–760.

36 Klinman DM, Takeshita F, Gursel I, Leifer C, Ishii KJ, Verthelyi D, Gursel M: CpG DNA: Recognition by and activation of monocytes. Microbes Infect 2002;4:897–901.

37 Hemmi H, Kaisho T, Takeda K, Akira S: The roles of Toll-like receptor 9, MyD88, and DNA-dependent protein kinase catalytic subunit in the effects of two distinct CpG DNAs on dendritic cell subsets. J Immunol 2003;170:3059–3064.

Shizuo Akira
Department of Host Defense, Research Institute for Microbial Diseases
Osaka University, 3–1 Yamada-oka, Suita, Osaka 565-0871 (Japan)
Tel. +81 6 6879 8303, Fax +81 6 6879 8305, E-Mail sakira@biken.osaka-u.ac.jp

Kabelitz D, Schröder J-M (eds): Mechanisms of Epithelial Defense.
Chem Immunol Allergy. Basel, Karger, 2005, vol 86, pp 136–150

Immunosurveillance by γδ T Cells: Focus on the Murine System

Michael Girardi[a], *Adrian C. Hayday*[b]

[a]Department of Dermatology, Yale University School of Medicine, New Haven, Conn., USA; [b]Peter Gorer Department of Immunobiology, Guy's King's St. Thomas' School of Medicine, Guy's Hospital, London, UK

Abstract

Physiologic immune responses are an integration of the activities of several lymphoid cell types that include qualitatively distinct T cell subsets. The contributions that specific T cell subsets make during infection, inflammation, and carcinogenesis is becoming increasingly clear from a variety of mouse models and, importantly, their backcrossing onto different genetic backgrounds. This review considers what we have learned in the mouse about the crucial roles played by γδ T cells. We consider how the cells' associations with specific tissues have revealed that T cell responses are regulated locally as well as systemically, and we discuss the implications of this for understanding and enhancing immune surveillance in the clinical setting.

TCRγδ$^+$ Intraepithelial Lymphocytes

Situated at the critical interface of the host organism with its environment, various epithelial surfaces are intermittently subjected to the damaging effects of microbial pathogens, chemical toxins, and mutagens. Physiologic defense against this requires unrelenting rounds of apoptosis, wound healing, and cell proliferation, but to prevent the systemic dissemination of infected or malignant cells, there is also a requirement for an immunological defense. This defense is based on the coordinated activities of various host immune cell populations that must balance the elimination of infected, metabolically stressed, and pre-malignant cells, with the limitation of a potentially over-exuberant response that might otherwise disrupt the epithelial barrier, paradoxically enhancing cellular damage and transformation. Central to these coordinated activities are intraepithelial lymphocytes (IELs), which primarily are T cells residing within body surfaces, where they sit interspersed with epithelial cells. Data derived from several laboratories collectively depict IELs as local T cells that can act relatively quickly and vigorously in response to infected and/or damaged epithelial cells, while also providing an immunoregulatory function [1, 2] (fig. 1).

Intriguingly, in humans and mice, IELs can be distinguished from systemic cells (of the blood, spleen and lymph nodes) not just by their anatomical location, but by their T cell receptor (TCR) expression. Whereas 90–99% of 'conventional' systemic T cells express highly diverse heterodimeric $\alpha\beta$ TCRs, IEL repertoires are enriched in oligoclonal cells expressing TCR$\gamma\delta$. In the human gut, TCR$\gamma\delta^+$ cells account for 10–25% of IELs, while in the murine skin they account for $\sim$100% of the resident IELs, commonly known as dendritic epidermal T cells (DETC) [3–5]. Moreover, DETC predominantly express an identical Vγ5Vδ1$^+$ TCR, while other IEL compartments are also either monoclonal or oligoclonal, expressing distinct albeit related TCRs (table 1).

Studies of fetal thymic development have shown that such signatory subsets of $\gamma\delta$ T cells emerge from the thymus in a series of development 'waves', guided to their tissues by the loss and gain of appropriate chemokine receptors [4, 6]. Thus, Vγ5Vδ1$^+$ thymocytes are the precursors of DETC, and are the first mature T cells to leave the mouse fetal thymus, whereupon they take up residence within the basal layer of the epidermis to form a dendritic network unique among T cells, but reminiscent of the skin's antigen-presenting Langerhans cells which are situated in a slightly more superficial epidermal plane [7].

Several studies have indicated that the particular TCR expressed by an IEL subset stabilizes its association with a particular tissue. For example, developing IEL compartments in the peri-neonatal period show a more diverse set of $\gamma\delta$ TCRs than is evident at 2–3 weeks of age [8]; intestinal IELs ectopically expressing a transgenic $\gamma\delta$ TCR not normally associated with the gut fail to mature properly [9]; and skin IEL repertoires from TCR$\delta^{-/-}$ mice, that per force express TCR$\alpha\beta$, seem to decay with time [10]. Such data are consistent with the hypothesis that signatory, tissue-associated $\gamma\delta$ TCRs recognize autologous, tissue-specific ligands expressed by epithelial cells in response to cell stress. Candidates for such ligands include inducible MHC Class IB gene products (discussed below) [11–14].

The proposed recognition of generic 'stress antigens' clearly distinguishes IELs from conventional systemic T cells that are highly specific for pathogen-encoded determinants that reflect the origin of an infection. Following the

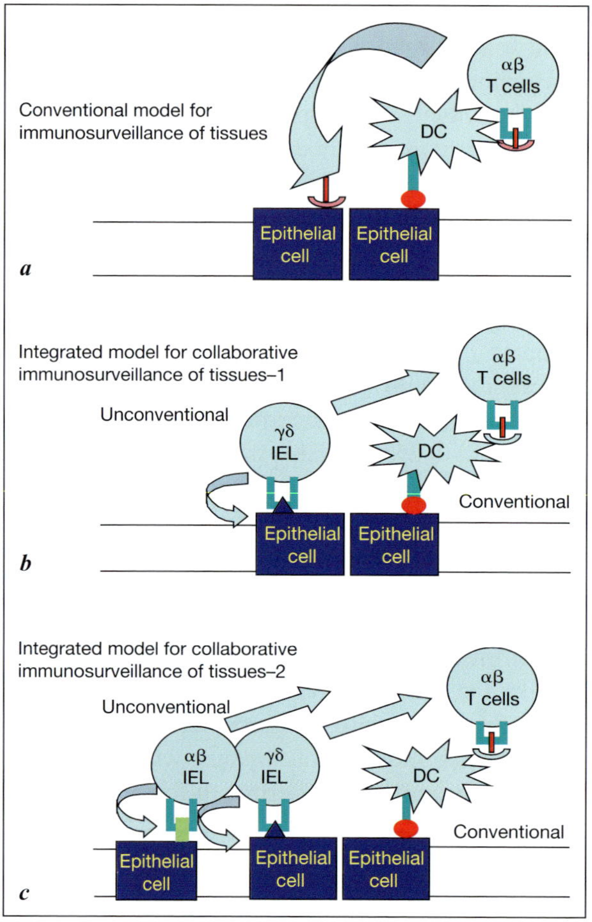

Fig. 1. Conventional and unconventional views of immunosurveillance. Illustrated is the transition from a conventional view of tissue immunosurveillance, to one in which unconventional T cells play a critical role. ***a*** In the conventional depiction, systemic αβ T cells, primed by dendritic cells (DC), return to infected or damaged epithelial surfaces where they recognize and respond to epithelial cells in an MHC dependent fashion. ***b*** The first unconventional depiction recognizes the capacity of tissue-associated γδ T cells to respond directly to stressed epithelial cells and to regulate the systemic infiltrates that home to the damaged tissues. ***c*** The second unconventional depiction recognizes that the roles played by tissue-associated γδ T cells may also be played by unconventional αβ T cells that share a similar genetic expression profile.

Table 1. TCR$\gamma\delta$ usage with murine epithelia

Murine tissue	TCR	T cell representation
Epidermis	Vγ5Vδ1	~100%
Uterine epithelium	Vγ6Vδ1	~100%
Small intestinal epithelium	Vγ7 with Vδ4/Vδ5	35–70%

recognition of 'stress antigens', it has been proposed that IELs kill the stressed epithelial cells and promote wound healing, as a result of which the potentially disastrous systemic dissemination of infected or transformed cells is prevented, while the integrity of the epithelium is maintained. The detailed analysis of IELs and of a number of transgenic and mutant mice has allowed this hypothesis to be tested.

Immunoprotective Roles of $\gamma\delta$ T Cells in the Tissues

TCR$\delta^{-/-}$ mice are deficient in all $\gamma\delta$ T cells. Inevitably, therefore, they have proven an invaluable tool for assessing the contributions of $\gamma\delta$ T cells to host responses. Nonetheless, before reviewing the information that the TCR$\delta^{-/-}$ mouse has provided, one must consider three caveats: first, not all $\gamma\delta$ T cells reside within the IEL compartments, and it is therefore inaccurate to assume that deficiencies shown by the TCR$\delta^{-/-}$ mouse reflect only IEL defects. Second (as discussed below), the TCR$\delta^{-/-}$ mouse is not deficient in all IELs. And third, the loss of a T cell subset, as is the case in the TCR$\delta^{-/-}$ mouse, may have led indirectly to defects in other cell types that may be responsible for any phenotypes observed [15].

This notwithstanding, when challenged with a variety of infectious agents (e.g. *Listeria*, vesicular stomatitis virus, malaria), TCR$\delta^{-/-}$ mice show enhanced susceptibility particularly at early time points after infection, consistent with the notion that $\gamma\delta$ T cells contribute to rapidly-mounted protective responses *via* recognition of common, stress-induced self-antigens [11]. In some instances, the contribution of $\gamma\delta$ cells to protection is very striking. For example, TCR$\delta^{-/-}$ mice infected with the lung pathogen *Nocardia asteroides* failed to mount a characteristic neutrophil attack, as a result of which there was unimpeded bacterial growth and epithelial necrosis [16]. Such data point to an important role for $\gamma\delta$ cells in coordinating early, innate responses. This is very different to the development of antigen-specific memory that is a signature of conventional $\alpha\beta$ T cells, a difference further highlighted by the finding that γδ cells likewise limit widespread lung damage in response to a noninfectious scenario, ozone exposure [16].

The perspective that γδ cells provide primary immunoprotection, rather than establishing antigen-specific memory, received further support from two independent studies of the responses of young mice to two related apicomplexan parasites, *Cryptosporidium* and the intestinal parasite *Eimeria vermiformis* [17, 18]. These studies showed that whereas adult mice mount robust antigen-specific αβ T cell responses to these pathogens, the protection of young mice relies heavily on the γδ cell compartment; results that are perhaps related to the early ontogenetic development of γδ T cell subsets [17, 18]. Nonetheless, there was no evidence for the development of antigen-specific memory in the γδ cell compartment [18].

Immunoregulatory Roles of Local γδ T Cells

Consistent with their contribution to immunoprotection, γδ T cells, particularly those in the IEL compartments, have been shown to be cytolytic [19–21]; to express various cytolytic mediators, e.g. granzymes and FasL [22–24]; and to be potent producers of cytokines such as IFNγ [25, 26]. At the same time, however, both focused and open-ended analyses of genes encoded by γδ T cells have revealed overt expression of a spectrum of chemokines and of RNAs encoding for other putative regulatory molecules [24, 27–29].

The finding that γδ T cells express RNAs for chemokines and other putative regulatory molecules appears consistent with several findings that immune responses to pathogens are not properly regulated in either TCRδ$^{-/-}$ mice or wild type mice in which γδ T cells have been depleted by antibody administration. As an example of this, adult TCRδ$^{-/-}$ mice infected with *Eimeria* showed exaggerated inflammatory responses, reflected in local hemorrhage and shedding of the intestinal epithelium. This phenomenon did not occur in animals deficient in both αβ and γδ T cells, suggesting that the γδ T cells were responsible for regulating the immunoprotective but tissue-damaging αβ T cell-driven inflammatory response [30]. Similar observations were made in several other models of infection including *Listeria monocytogenes*, *Mycobacterium tuberculosis*, and *Klebsiella* [31–34]. For example, mice infected with *L. monocytogenes* in a single testicle develop a contralateral inflammation in the other testicle, which is substantially exaggerated in mice lacking γδ cells [32].

TCRδ$^{-/-}$ mice have also shown accelerated immune responses in noninfectious settings. For example, the αβ T cell-driven, lupus-like disease of MRL/*lpr* mice was markedly worsened when the mice were rendered deficient in γδ T cells by backcrossing the TCRδ$^{-/-}$ mutation onto the susceptible background; whereas ~25% of MRL/*lpr* mice die (primarily of glomerulonephritis) after

6 months, this figure rises to ~67% for TCRδ$^{-/-}$ MRL/*lpr* mice [35]. Roughly contemporary with these studies, Shiohara and colleagues observed that γδ cells would regulate αβ T cell activity in the skin [36].

In their studies, Shiohara and colleagues locally injected mice with αβ T cells auto-reactive to Langerhans cells (LC) in the epidermis. After the LC were depleted, and the auto-reactive αβ T cells (with no stem cell potential) had died off, the mice recovered, displayed increased numbers of DETC, and were resistant to subsequent challenges with the autoreactive T cells. By contrast, TCRδ$^{-/-}$ mice injected in the same fashion showed an exaggerated epidermal infiltration, and failed to develop resistance to re-challenge [36]. Repopulation of the TCRδ$^{-/-}$ mice with γδ T cells prior to challenge restored the wild type phenotype, revealing the capacity of γδ T cells to regulate αβ T cell-mediated immunity in the skin. Nevertheless, two fundamental questions arising from these studies remained to be elucidated: (1) the degree to which γδ T cells were the natural, physiologic mediators of immunoregulation (as opposed to immunoregulation simply being a function that *could* be effected by γδ T cells under particular, contrived circumstances), and (2) whether the locally resident TCRγδ$^+$ DETC, in the absence of other (e.g. systemic recirculating) TCRγδ$^+$ T cells, are sufficient to protect the epidermis against exaggerated αβ T cell-mediated inflammation. These issues were addressed by a set of studies in which the TCRδ$^{-/-}$ mutation was backcrossed onto other genetic backgrounds, notably NOD and FVB [37].

Essentially 100% of TCRδ$^{-/-}$ NOD mice show a spontaneous dermatitis, most evident in the unprotected ear skin, despite housing under specific pathogen-free conditions. Likewise, TCRδ$^{-/-}$ FVB mice consistently show exaggerated cutaneous inflammatory responses to contact allergens and irritants. As is the case for γδ T cell immunoregulation in the *Eimeria* model and the MRL/lpr model, the spontaneous and augmented contact responses in γδ deficient mice on susceptible backgrounds are dependent on αβ T cells; that is, the phenotype is absent in double-knockout (TCRβ$^{-/-}$δ$^{-/-}$) mice which lack both αβ and γδ T cells [37, and unpubl. obs.]. These results, and particularly the spontaneous development of pathology in TCRδ$^{-/-}$ NOD mice, show that γδ cells are natural physiologic regulators of cutaneous inflammation, and that their role in the mouse is nonredundant with that of other cells.

When susceptible γδ-deficient mice were injected at birth with DETC progenitors (Vγ5$^+$ thymocytes obtained from fetal thymi), their epidermis was selectively repopulated with Vγ5$^+$ DETC, and they did not develop spontaneous or augmented induced dermatitis. The same result was not achieved when mice were repopulated with systemic γδ cells [37]. These studies unequivocally established that cutaneous immunoregulation was effected by local TCRγδ$^+$ IELs, logically suggesting that local T cells might also regulate systemic responses in other sites, e.g. the intestinal epithelium, a notion consistent with reports that IEL deficiencies are associated with human inflammatory bowel disease pathologies [38]. Another clear and important finding of these studies was that the regulation of αβ T cell responses by γδ cells occurred in the effector phase of the response [37]. That is to say, the lack of γδ cells was not associated with enhanced priming and increased numbers of responding αβ T cells in the secondary lymphoid tissues, but rather was associated with an exaggerated effect of those cells in the tissue to which they migrated.

Although developing spontaneously, the dermatitis that develops in $TCR\delta^{-/-}$ NOD mice can be reduced by housing mice in individually ventilated caging where humidity is low and there is less accumulation of chemical irritants. This strong environmental influence on αβ T cell-driven dermatitis shares several properties with the human disorder, atopic dermatitis (AD) [2, 39]. In addition to similar histologic features, they both show compromised barrier function, exacerbation with chemical irritants, and augmented responses to contact allergens. Moreover, they are both strikingly influenced by genetic modifiers: thus, $TCR\delta^{-/-}$ C57BL/6 mice show little or no difference in cutaneous inflammation by comparison to their cage-matched $TCR\delta^{+}$ controls. This observation may be exploited *via* genetic mapping to identify genes that regulate the development of dermatitis that occurs in the absence of γδ T cells: such genes may have broad implications for the local control of cutaneous inflammation and for AD.

Finally, an important aspect of the gene expression data that have been obtained for γδ cells is the likely pleiotropy of γδ cells. Indeed, the capacity of γδ cells to limit chronic lymphocytic inflammation should be set against the cells' capacity to promote neutrophil attack in the lungs in response to *Nocardia* infection (see above). It is therefore unreasonable to assume that γδ cells are simply anti-inflammatory in all situations: instead, they should be viewed as cells with potent capacity to regulate systemic lymphoid responses, with the nature of the regulation most likely being specific to the tissue concerned and/or the broad nature of the challenge. Indeed, O'Brien and colleagues studying the γδ cell response to enterovirus infection of mice concluded that γδ cells expressing $V\gamma1^{+}$ TCRs were anti-inflammatory while those with $V\gamma4^{+}$ TCRs were proinflammatory [40].

$\gamma\delta^{+}$ T Cells and Tumor Surveillance

In human disease there is a strong association of chronic inflammatory states with malignant transformation, e.g. intestinal malignancy (with inflammatory bowel disease), and cutaneous squamous cell carcinoma (with chronic ulcers). Because of the increased inflammation that occurs in particular strains of $TCR\delta^{-/-}$

mice, it was natural to ask whether TCR$\delta^{-/-}$ mice show increased susceptibility to malignancy. Indeed, it is provocative that one of the backgrounds, FVB, that develops exaggerated cutaneous inflammation in the absence of γδ cells is a strain that is highly susceptible to chemical carcinogenesis.

Most cancer immunotherapy strategies have focused on the stimulation of anti-tumor αβ T cells via vaccination of patients with tumor antigens and/or peptides, often pre-loaded onto dendritic cells [41, 42]. While it is feasible to identify and stimulate tumor-antigen-specific responses, especially for malignancies with relatively high immunogenicity (e.g. melanoma), the tumor cell population does not remain static in the face of the anti-tumor force exerted by αβ T cells [43]. Highly proliferative and with unstable genomes, tumor populations become edited, with concurrent selection for those cells that have altered their expression of tumor antigens and/or the molecules that present them [43]. Moreover, tumor cells may evolve to actively suppress effector αβ T cells, e.g. *via* secretion of immunoregulatory cytokines; the expression of FasL that can kill activated T cells; or the expression of serpins that confer tumor cells resistance to granzyme-mediated attack. Therefore, the full potential of immunotherapy is unlikely to be realized by strategies that focus only on αβ T cells. Instead, a broader understanding of the full complexion of the antitumor response is required.

A number of studies in mice have pointed to the capacity of γδ cells to regulate malignancy. Almost a decade ago, Mak and colleagues noted that mice expressing a TCRVγ1 chain transgene were spontaneously resistant to acute T-cell leukemias although they did not reject non-hematopoietic tumors [44]. By comparison to IELs, TCRV$\gamma 1^+$ cells are most often found in the lymphoid organs and blood, and TCRV$\gamma 1^+$ hybridomas isolated from the transgenic mice reacted in vitro against almost all hematopoietic tumor cell lines tested, in a fashion independent of MHC class I, MHC class II, or the TAP peptide transporter [44]. Shortly thereafter, it was shown that *lpr* mice with mutations in the Fas gene rapidly and spontaneously developed massive splenic B cell malignancies if the mice lacked both αβ T cells and γδ cells, but that this was not the case if either T cell subset were present [45]. Both sets of data point to a capacity of γδ cells to regulate the development of hematologic malignancies, consistent with which human clinical trials are now ongoing in which the enhanced activities of γδ cells are being harnessed against non-Hodgkin's lymphoma and multiple myeloma [46].

γδ$^+$ T Cell Regulation of Epithelial Malignancy

More recently, the contributions of γδ cells to cutaneous tumor surveillance have been systematically studied. Intradermal injection of the chemical mutagen methylcholanthrene (MCA) can provoke the development of poorly differentiated fibrosarcomas, and spindle cell carcinomas. This occurs with greater frequency and shorter latency in TCR$\delta^{-/-}$ mice [47, 48]. Similarly, tumors form more readily in TCR$\delta^{-/-}$ mice following injection of squamous cell carcinoma (SCC) or melanoma tumor cell lines [47, 48]. Suggesting that γδ T cells regulate carcinogenesis more broadly, chemically induced colorectal adenocarcinoma reportedly occurs with greater frequency in γδ-deficient mice [49]. Although the mechanisms by which γδ cells exert their effects are incompletely understood, there are numerous instances in which TCR$\gamma\delta^+$ IELs have been shown to directly kill transformed and/or stressed epithelial cell targets, again in an MHC-independent fashion [47, 50]. Additionally, their activation-induced capacity to make IFNγ in the early stages of tumor growth appears to be important [48]. The notion that the different T cell compartments may contribute to tumor surveillance in distinct fashions and at distinct stages of tumor growth has been further explored in a two-stage system of chemical carcinogenesis. wherein a single epicutaneous application of dimethylbenzanthracine (DMBA) is followed by repeated application of low concentrations of the tumor promoting agent, TPA [51].

The two-stage protocol permits one to measure two features of carcinogenesis: the incidence of *tumor development* via monitoring of papilloma formation, and the frequency of *tumor progression* via quantitation of the conversion of papillomas into frank SCC. In TCR$\delta^{-/-}$ FVB mice, the appearance of papillomas and the progression to carcinoma were both significantly increased [41, 52]. These data are consistent with the proposal that γδ T cells can inhibit the early stages of tumor development (see above), but that they also limit progression to carcinoma [52]. One may hypothesize that these respective effects of γδ cells may reflect their utilization of two different mechanisms.

A clue to this was provided by the study of tumor incidence and progression in mice lacking αβ T cells [52]. In this case, there was a much less obvious increase in papilloma development. Moreover, under an intense regimen of chemical promotion, tumors were *less* likely to progress to the carcinoma stage than was the case in wild type mice [52]. This highlighted a paradoxical tumor-promoting effect attributable to αβ T cells, a concept that is consistent with other experimental systems where pro-tumor contributions of αβ T cells have been reported [53, 54]. Intriguingly, these results suggest that γδ cells might limit tumor progression by limiting the pro-inflammatory responses of αβ T cells, much as they have been described to do in several other systems (see above).

In sum, as is the case in the host response to infection and to physical stress, γδ cells may exert two qualitatively distinct effects: a rapidly-responsive immunoprotective effect (cytolysis; IFNγ, etc.) and a critical immunoregulatory mechanism. Although the latter is poorly understood, it is clear from murine systems that γδ cells can, in some circumstances, kill or otherwise dysregulate APCs, thereby limiting the αβ T cell response [55, 56]. Moreover, γδ T cell elimination of activated Fas^+ αβ T cells mediating Coxsackie B3 virus-induced mycocarditis has been implicated as a mechanism by which γδ T cells may again limit αβ T cell mediated immunopathology [57].

Another, not mutually exclusive, means by which γδ cells might reduce the infiltrative effects of systemic αβ T cells is to increase the resistance of the epithelial tissue itself. In this regard, $TCR\gamma\delta^+$ IELs have been reported to produce fibroblast growth factors 7 and 10 [27] as well as inhibin β [Wise and Tigelaar, in preparation], and to enhance wound healing in the skin and the gut [58, 59].

From γδ Cell Biology in Mice to Immunosurveillance in Humans

To what degree do the mouse studies inform us of immunoprotection and immunoregulation in humans? A conspicuous obstacle to direct extrapolation is the fact that there is no obvious homologue of murine (or bovine) DETC in the human epidermis. We consider several responses to this seeming paradox. Perhaps the most simple is to invoke an analogous γδ T cell population, albeit with different characteristics. In this regard, Holtmeier et al. [60] have provided clear evidence for human dermal γδ cells displaying a TCR repertoire that could be distinguished from that of systemic γδ cells. There is likewise a human, gut-associated $TCR\gamma\delta^+$ IEL compartment that is distinct from systemic γδ cells. Clearly, the functional potential of these cells needs to be thoroughly clarified. (For further elucidation on this subject, the reader is referred to the review by Holtmeier and Kabelitz, entitled 'Gammadelta T cells link innate and adaptive immune responses. Focus on human Vγ9/Vδ2 and Vδ1 T cells,' immediately following this review [pp. 151–183].)

A second possibility is that the crucial mechanisms exerted by murine DETC are effected by other, non-$TCR\gamma\delta^+$ cells in the human. Support for this hypothesis comes from the finding that γδ cells are unlikely to be unique even in the mouse. A comprehensive comparison of murine gut associated $TCR\gamma\delta^+$ and $TCR\alpha\beta^+$ IELs identified numerous genes expressed substantially more strongly by the latter cells, but essentially no γδ-specific genes. It was hypothesized that the $TCR\alpha\beta^+$ population may contain within it a set of 'unconventional' $TCR\alpha\beta^+$ cells that very closely resembled $TCR\gamma\delta^+$ cells [24, 61].

In support of this hypothesis, the mouse intestine harbors substantial numbers of $CD8\alpha\alpha^+$ $TCR\alpha\beta^+$ IEL that resemble γδ IEL in their limited TCR diversity, and expression of strikingly similar gene profiles [15]. Moreover, Poussier et al. [62] showed that CD8$\alpha\alpha^+$ TCR$\alpha\beta^+$ cells regulate systemic $\alpha\beta$ T cell-driven colitis. Hence, $\gamma\delta$ cells may be mere prototypes of unconventional T cells that in different tissues in different species play important roles in the rapid response to challenge and in the regulation of systemic responses (fig. 1) [15, 61]. This perspective is consistent with the observations that the most overt phenotypes of TCR$\delta^{-/-}$ occur in the skin, where there are few if any $\alpha\beta$ T cell populations that might compensate for their loss. In the gut, by contrast, unconventional $\alpha\beta$ T cells may contribute to defense and immunoregulation in the absence of $\gamma\delta\delta$ cells.

As a summation of a cell's activity, rapid primary protection and immunoregulation could be descriptors of natural killer (NK) and/or NKT cells, in which regard it is striking that the gene profiles of unconventional T cells include many genes, such as LAG3, DAP12, NKG2D, and granzyme C that were hitherto primarily associated with NK cells [15]. Hence, one must consider the possibility that the human equivalents of tissue-associated murine $\gamma\delta$ cells may include NK cells.

Immunological Mechanisms Highlighted by $\gamma\delta$ Cells – Towards the Clinic

One approach to identifying which human cells are the functional equivalents of murine $\gamma\delta$ cells is to identify the cells that share signatory molecular mechanisms with $\gamma\delta$ cells. For example, DETC constitutively express the activating receptor, NKG2D, that recognizes MHC class IB molecules, such as Rae-1 and H60, that are induced in the skin by chemical carcinogens [47, 63–65]. Human NKG2D likewise recognizes the stress-induced MHC class IB molecules, MICA and MICB, which are expressed by myriad solid tumors [65, 66].

The activation of cells via NKG2D is increasingly recognized as an important component both innate and adaptive immunosurveillance, and evidence of its contribution to the biology of DETC was provided by the finding that the targeting of SCC cells in vitro by DETC was inhibitable with anti-NKG2D antibodies. Moreover, the avoidance of recognition by NKG2D$^+$ cells is an overt mechanism of immuno-evasion in mice and in human [67, 68; Oppenheim, submitted]. It is therefore important to identify in humans the NKG2D$^+$ cells that may be mounting an analogous response to transformed cells. While the situation in the skin is unclear, it is evident that many gut-associated solid tumors are infiltrated by NKG2D$^+$ TCR$\gamma\delta^+$ IELs. In parallel, the clarification of the tumor-associated ligands recognized directly by TCR$\gamma\delta$

will permit the identification of the cells that respond to the equivalent ligands on human carcinomas [69, 70].

In summary, animal models have proved a powerful test-bed for the hypothesis that IELs are activated by stress-induced antigens, whereupon they provide rapid primary protection, and protect epithelial integrity against the immunopathologic potential of systemic infiltration. Indeed, the animal studies have exposed critical roles for TCR$\gamma\delta^+$ cells, including those in the IEL compartment, in inflammation and cancer biology. The studies have also identified other unconventional lymphocytes subsets with similar functional properties. The increasing capacity to identify precisely which molecules are used by such unconventional lymphocytes to exert their effects provides a clear path to identifying molecules and pathways that may have application in the clinical treatment of inflammatory disease, cancer, and other forms of tissue disruption.

References

1 Girardi M: Cutaneous Biology of γδ T Cells. Adv Dermatol 2004;20:203–215.

2 Hayday A, Tigelaar R: Immunoregulation in the tissues by gammadelta T cells. Nat Rev Immunol 2003;3:233–242.

3 Asarnow DM, Goodman T, LeFrancois L, Allison JP: Distinct antigen receptor repertoires of two classes of murine epithelium-associated T cells. Nature 1989;341:60–62.

4 Allison JP, Havran WL: The immunobiology of T cells with invariant gamma delta antigen receptors. Annu Rev Immunol 1991;9:679–705.

5 Goodman T, Lefrancois L: Expression of the gamma-delta T-cell receptor on intestinal CD8$^+$ intraepithelial lymphocytes. Nature 1988;333:855–858.

6 Itohara S, Farr AG, Lafaille JJ, Bonneville M, Takagaki Y, Haas W, Tonegawa S: Homing of a gamma delta thymocyte subset with homogeneous T-cell receptors to mucosal epithelia. Nature 1990;343:754–757.

7 Bergstresser PR, Sullivan S, Streilein JW, Tigelaar RE: Origin and function of Thy-1+ dendritic epidermal cells in mice. J Invest Dermatol 1985;85(suppl 1):85s–90s.

8 Kyes S, Pao W, Hayday A: Influence of site of expression on the fetal gamma delta T-cell receptor repertoire. Proc Natl Acad Sci USA 1991;88:7830–7833.

9 Bonneville M, Itohara S, Krecko EG, Mombaerts P, Ishida I, Katsuki M, Berns A, Farr AG, Janeway CA Jr, Tonegawa S: Transgenic mice demonstrate that epithelial homing of gamma/delta T cells is determined by cell lineages independent of T cell receptor specificity. J Exp Med 1990;171:1015–1026.

10 Jameson JM, Cauvi G, Witherden DA, Havran WL: A keratinocyte-responsive gamma delta TCR is necessary for dendritic epidermal T cell activation by damaged keratinocytes and maintenance in the epidermis. J Immunol 2004;172:3573–3579.

11 Janeway CA Jr, Jones B, Hayday A: Specificity and function of T cells bearing gamma delta receptors. Immunol Today 1988;9:73–76.

12 Chien YH, Jores R, Crowley MP: Recognition by γδ T cells. Annu Rev Immunol 1996;14: 511–532.

13 Steele CR, Oppenheim DE, Hayday AC: Gamma/delta T cells: Non-classical ligands for non-classical cells. Curr Biol 2000;10:R282–285.

14 Wu J, Groh V, Spies T: T cell antigen receptor engagement and specificity in the recognition of stress-inducible MHC class I-related chains by human epithelial gamma delta T cells. J Immunol 2002;169:1236–1240.

15 Pennington DJ, Silva-Santos B, Shires J, Theodoridis E, Pollitt C, Wise EL, Tigelaar RE, Owen MJ, Hayday AC: The inter-relatedness and interdependence of mouse T cell receptor gammadelta$^+$ and alphabeta$^+$ cells. Nat Immunol 2003;4:991–998.
16 King DP, Hyde DM, Jackson KA, Novosad DM, Ellis TN, Putney L, Stovall MY, Van Winkle LS, Beaman BL, Ferrick DA: Cutting edge: Protective response to pulmonary injury requires gamma delta T lymphocytes. J Immunol 1999;162:5033–5036.
17 Waters WR, Harp JA: *Cryptosporidium parvum* infection in T-cell receptor (TCR)-alpha- and TCR-delta-deficient mice. Infect Immun 1996;64:1854–1857.
18 Ramsburg E, Tigelaar R, Craft J, Hayday A: Age-dependent requirement for gammadelta T cells in the primary but not secondary protective immune response against an intestinal parasite. J Exp Med 2003;198:1403–1414.
19 Kaminski MJ, Cruz PD Jr, Bergstresser PR, Takashima A: Killing of skin-derived tumor cells by mouse dendritic epidermal T-cells. Cancer Res 1993;53:4014–4019.
20 Havran WL, Poenie M, Tigelaar RE, Tsien RY, Allison JP: Phenotypic and functional analysis of gamma delta T cell receptor-positive murine dendritic epidermal clones. J Immunol 1989;142: 1422–1428.
21 Guy-Grand D, Malassis-Seris M, Briottet C, Vassalli P: Cytotoxic differentiation of mouse gut thymodependent and independent intraepithelial T lymphocytes is induced locally. Correlation between functional assays, presence of perforin and granzyme transcripts, and cytoplasmic granules. J Exp Med 1991;173:1549–1552.
22 Mohamadzadeh M, McGuire MJ, Smith DJ, Gaspari AA, Bergstresser PR, Takashima A: Functional roles for granzymes in murine epidermal gamma(delta) T-cell-mediated killing of tumor targets. J Invest Dermatol 1996;107:738–742.
23 Krahenbuhl O, Gattesco S, Tschopp J: Murine Thy-1$^+$ dendritic epidermal T cell lines express granule-associated perforin and a family of granzyme molecules. Immunobiology 1992;184:392–401.
24 Shires J, Theodoridis E, Hayday AC: Biological insights into TCRgamma delta$^+$ and TCRalpha beta$^+$ intraepithelial lymphocytes provided by serial analysis of gene expression (SAGE). Immunity 2001;15:419–434.
25 Huber H, Descossy P, Regier E, van Brandwijk R, Knop J: Activation of phenotypically heterogeneous murine T cell receptor gamma delta$^+$ dendritic epidermal T cells by self-antigen(s). Int Arch Allergy Immunol 1995;107:498–507.
26 Matsue H, Cruz PD Jr, Bergstresser PR, Takashima A: Profiles of cytokine mRNA expressed by dendritic epidermal T cells in mice. J Invest Dermatol 1993;101:537–542.
27 Boismenu R, Havran WL: Modulation of epithelial cell growth by intraepithelial gamma delta T cells. Science 1994;266:1253–1255.
28 Boismenu R, Feng L, Xia YY, Chang JC, Havran WL: Chemokine expression by intraepithelial gamma delta T cells. Implications for the recruitment of inflammatory cells to damaged epithelia. J Immunol 1996;157:985–992.
29 Fahrer AM, Konigshofer Y, Kerr EM, Ghandour G, Mack DH, Davis MM, Chien YH: Attributes of gammadelta intraepithelial lymphocytes as suggested by their transcriptional profile. Proc Natl Acad Sci USA 2001;98:10261–10266.
30 Roberts SJ, Smith AL, West AB, Wen L, Findly RC, Owen MJ, Hayday AC: T-cell alpha beta$^+$ and gamma delta$^+$ deficient mice display abnormal but distinct phenotypes toward a natural, widespread infection of the intestinal epithelium. Proc Natl Acad Sci USA 1996;93:11774–11779.
31 Mombaerts P, Arnoldi J, Russ F, Tonegawa S, Kaufmann SH: Different roles of alpha beta and gamma delta T cells in immunity against an intracellular bacterial pathogen. Nature 1993;365:53–56.
32 Mukasa A, Hiromatsu K, Matsuzaki G, O'Brien R, Born W, Nomoto K: Bacterial infection of the testis leading to autoaggressive immunity triggers apparently opposed responses of alpha beta and gamma delta T cells. J Immunol 1995;155:2047–2056.
33 D'Souza CD, Cooper AM, Frank AA, Mazzaccaro RJ, Bloom BR, Orme IM: An anti-inflammatory role for gamma delta T lymphocytes in acquired immunity to *Mycobacterium tuberculosis*. J Immunol 1997;158:1217–1221.
34 Moore TA, Moore BB, Newstead MW, Standiford TJ: Gamma delta-T cells are critical for survival and early proinflammatory cytokine gene expression during murine Klebsiella pneumonia. J Immunol 2000;165:2643–2650.

35 Peng SL, Madaio MP, Hayday AC, Craft J: Propagation and regulation of systemic autoimmunity by gammadelta T cells. J Immunol 1996;157:5689–5698.

36 Shiohara T, Moriya N, Hayakawa J, Itohara S, Ishikawa H: Resistance to cutaneous graft-vs.-host disease is not induced in T cell receptor delta gene-mutant mice. J Exp Med 1996;183:1483–1489.

37 Girardi M, Lewis J, Glusac E, Filler RB, Geng L, Hayday AC, Tigelaar RE: Resident skin-specific gammadelta T cells provide local, nonredundant regulation of cutaneous inflammation. J Exp Med 2002;195:855–867.

38 Van Damme N, De Keyser F, Demetter P, Baeten D, Mielants H, Verbruggen G, Cuvelier C, Veys EM, DeVos M: The proportion of Th1 cells, which prevail in gut mucosa, is decreased in inflammatory bowel syndrome. Clin Exp Immunol 2001;125:383–390.

39 Beltrani VS: The clinical spectrum of atopic dermatitis. J Allergy Clin Immunol 1999;104: S87–98.

40 Huber SA, Graveline D, Newell MK, Born WK, O'Brien RL: $V\gamma1^+$ T cells suppress and $V\gamma4^+$ T cells promote susceptibility to coxsackievirus B3-induced myocarditis in mice. J Immunol 2000;165:4174–4181.

41 Coulie PG, Karanikas V, Colau D, Lurquin C, Landry C, Marchand M, Dorval T, Brichard V, Boon T: A monoclonal cytolytic T-lymphocyte response observed in a melanoma patient vaccinated with a tumor-specific antigenic peptide encoded by gene MAGE-3. Proc Natl Acad Sci USA 2001; 98:10290–10295.

42 Sivinski CL, Kohlgraf KG, VanLith ML, Morikane K, Tempero RM, Hollingsworth MA: Molecular requirements for CD8-mediated rejection of a MUC1-expressing pancreatic carcinoma: Implications for tumor vaccines. Cancer Immunol Immunother 2002;51:327–340.

43 Dunn GP, Old LJ, Schreiber RD The immunobiology of cancer immunosurveillance and immunoediting. Immunity 2004;21:137–148.

44 Penninger JM, Wen T, Timms E, Potter J, Wallace VA, Matsuyama T, Ferrick D, Sydora B, Kronenberg M, Mak TW: Spontaneous resistance to acute T-cell leukaemias in TCRVγ1.1Jγ4Cγ4 transgenic mice. Nature 1995;375:241–244.

45 Peng SL, Robert ME, Hayday AC, Craft J: A tumor-suppressor function for Fas (CD95) revealed in T cell-deficient mice. J Exp Med 1996;184:1149–1154.

46 Wilhelm M, Kunzmann V, Eckstein S, Reimer P, Weissinger F, Ruediger T, Tony HP: $\gamma\delta$ T cells for immune therapy of patients with lymphoid malignancies. Blood 2003;102:200–206.

47 Girardi M, Oppenheim DE, Steele CR, Lewis JM, Glusac E, Filler R, Hobby P, Sutton B, Tigelaar RE, Hayday AC: Regulation of cutaneous malignancy by $\gamma\delta$ T cells. Science 2001;294:605–609.

48 Gao Y, Yang W, Pan M, Scully E, Girardi M, Augenlicht LH, Craft J, Yin Z: $\gamma\delta$ T cells provide an early source of interferon gamma in tumor immunity. J Exp Med 2003;198:433–442.

49 Matsuda S, Kudoh S, Katayama S: Enhanced formation of azoxymethane-induced colorectal adenocarcinoma in gammadelta T lymphocyte-deficient mice. Jpn J Cancer Res 2001;92:880–885.

50 Havran WL, Chien YH, Allison JP: Recognition of self antigens by skin-derived T cells with invariant gamma delta antigen receptors. Science 1991;252:1430–1432.

51 Owens DM, Wei S, Smart RC: A multihit, multistage model of chemical carcinogenesis. Carcinogenesis 1999;20:1837–1844.

52 Girardi M, Glusac E, Filler RB, Roberts SJ, Propperova I, Lewis J, Tigelaar RE, Hayday AC: The distinct contributions of murine T cell receptor (TCR)gamma delta$^+$ and TCRalpha beta$^+$ T cells to different stages of chemically induced skin cancer. J Exp Med 2003;198:747–755.

53 Siegel CT, Schreiber K, Meredith SC, Beck-Engeser GB, Lancki DW, Lazarski CA, Fu YX, Rowley DA, Schreiber H: Enhanced growth of primary tumors in cancer-prone mice after immunization against the mutant region of an inherited oncoprotein. J Exp Med 2000;191: 1945–1956.

54 Daniel D, Meyer-Morse N, Bergsland EK, Dehne K, Coussens LM, Hanahan D: Immune enhancement of skin carcinogenesis by $CD4^+$ T cells. J Exp Med 2003;197:1017–1028.

55 Egan PJ, Carding SR: Downmodulation of the inflammatory response to bacterial infection by gammadelta T cells cytotoxic for activated macrophages. J Exp Med 2000;191:2145–2158.

56 Skeen MJ, Freeman MM, Ziegler HK: Changes in peritoneal myeloid populations and their proinflammatory cytokine expression during infection with Listeria monocytogenes are altered in the absence of γ/δ T cells. J Leukoc Biol 2004;76:104–115.

57 Huber SA, Graveline D, Born WK, O'Brien RL: Cytokine production by Vgamma(+)-T-cell subsets is an important factor determining CD4(+)-Th-cell phenotype and susceptibility of BALB/c mice to coxsackievirus B3-induced myocarditis. J Virol 2001;75:5860–5869.
58 Jameson J, Ugarte K, Chen N, Yachi P, Fuchs E, Boismenu R, Havran WL: A role for skin γδ T cells in wound repair. Science 2002;296:747–749.
59 Chen Y, Chou K, Fuchs E, Havran WL, Boismenu R: Protection of the intestinal mucosa by intraepithelial γδ T cells. Proc Natl Acad Sci USA 2002;99:14338–14343.
60 Holtmeier W, Pfander M, Hennemann A, Zollner TM, Kaufmann R, Caspary WF: The TCR-δ repertoire in normal human skin is restricted and distinct from the TCR-δ repertoire in the peripheral blood. J Invest Dermatol 2001;116:275–280.
61 Hayday A, Theodoridis E, Ramsburg E, Shires J: Intraepithelial lymphocytes: Exploring the third way in immunology. Nat Immunol 2001;2:997–1003
62 Poussier P, Ning T, Banerjee D, Julius M: A unique subset of self-specific intraintestinal T cells maintains gut integrity. J Exp Med 2002;195:1491–1497.
63 Cerwenka A, Bakker AB, McClanahan T, Wagner J, Wu J, Phillips JH, Lanier LL: Retinoic acid early inducible genes define a ligand family for the activating NKG2D receptor in mice. Immunity 2000;12:721–727.
64 Diefenbach A, Jamieson AM, Liu SD, Shastri N, Raulet DH: Ligands for the murine NKG2D receptor: Expression by tumor cells and activation of NK cells and macrophages. Nat Immunol 2000;1:119–126.
65 Jamieson AM, Diefenbach A, McMahon CW, Xiong N, Carlyle JR, Raulet DH: The role of the NKG2D immunoreceptor in immune cell activation and natural killing. Immunity 2002;17:19–29.
66 Groh V, Rhinehart R, Secrist H, Bauer S, Grabstein KH, Spies T: Broad tumor-associated expression and recognition by tumor-derived gamma delta T cells of MICA and MICB. Proc Natl Acad Sci USA 1999;96:6879–6884.
67 Groh V, Wu J, Yee C, Spies T: Tumour-derived soluble MIC ligands impair expression of NKG2D and T-cell activation. Nature 2002;419:734–738.
68 Salih HR, Rammensee HG, Steinle A: Cutting edge: Down-regulation of MICA on human tumors by proteolytic shedding. J Immunol 2002;169:4098–4102.
69 Steele CR, Van Remoortere KC, Hayday AC: Production of a soluble gammadelta T-cell receptor to identify ligands for the murine intestinal intraepithelial gammadelta T cell population. J Chromatogr [B] 2003;786:297–304.
70 Aydintug MK, Roark CL, Yin X, Wands JM, Born WK, O'Brien RL: Detection of cell surface ligands for the γδ TCR using soluble TCRs. J Immunol 2004;172:4167–4175.

Michael Girardi, MD
Department of Dermatology, Yale University School of Medicine
333 Cedar Street, New Haven, CT 06520–8059 (USA)
Tel. +1 203 785 4092, Fax +1 203 785 7234, E-Mail michael.girardi@yale.edu

Kabelitz D, Schröder J-M (eds): Mechanisms of Epithelial Defense.
Chem Immunol Allergy. Basel, Karger, 2005, vol 86, pp 151–183

γδ T Cells Link Innate and Adaptive Immune Responses

Focus on Human Vγ9/Vδ2 and Vδ1 T Cells

Wolfgang Holtmeier[a,*], *Dieter Kabelitz*[b,**]

[a]Medizinische Klinik I., Johann Wolfgang Goethe-Universität, Frankfurt am Main, and [b]Institute of Immunology, UK S-H Campus Kiel, Kiel, Germany

Abstract

While most T cells use a CD3-associated α/β T cell receptor as antigen recognition structure, a second population of T cells expresses the alternative γ/δ T cell receptor. γ/δ T cells are a minor population in the peripheral blood but constitute a major population among intestinal intraepithelial lymphocytes. Most γ/δ T cells recognize ligands which are fundamentally different from the short peptides that are seen by α/β T cells in the context of MHC class I or class II molecules. Thus, human Vδ2 T cells recognize small bacterial phosphoantigens, alkylamines and synthetic aminobisphosphonates, whereas Vδ1 T cells recognize stress-inducible MHC-related molecules MICA/B as well as several other ligands. At the functional level, γ/δ T cells rapidly produce a variety of cytokines and usually exert potent cytotoxic activity, also towards many tumor cells. In this article, we discuss the role of γ/δ T cells as a bridge between the innate and the adaptive immune system, based on the interpretation that γ/δ T cells use their T cell receptor as a pattern recognition receptor. Our increasing understanding of the ligand recognition and activation mechanisms of γ/δ T cells also opens new perspectives for the development of γ/δ T cell-based immunotherapies.

Introduction

Two main subsets of γ/δ T cells have been described in humans, i.e. one expressing the TCR variable regions Vγ9/Vδ2 and the other expressing Vδ1 paired with various Vγ elements [1]. Recent evidence suggests that these

Supported by the Deutsche Forschungsgemeinschaft *Ho 1521/3–2, and **through SFB 617 (project A16).

subsets display unique features in terms of antigen specificities, requirements for antigen recognition and tissue distribution [2]. Vγ9/Vδ2 lymphocytes represent the majority of peripheral γ/δ T cells whereas Vδ1 lymphocytes are resident mainly within epithelial tissues. Vγ9/Vδ2 T cells recognize non-peptide antigens such as phosphate-containing compounds and alkylamines which can be found in a large variety of microbes and plants but also in mammalian cells. By contrast, Vδ1 T cells recognize different classes of antigens such as self lipids presented by CD1 and the stress inducible MHC class I-related chains A and B (MICA and MICB). It is thought that both γ/δ T cell subsets can provide natural resistance to microbial infections and perhaps tumors.

Vγ9/Vδ2 and Vδ1 T cells are part of the adaptive immune system in that they have a memory phenotype and junctionally diverse TCRs that require gene rearrangement. On the other hand, Vγ9/Vδ2 T cells also are part of the innate immune response, because they express a restricted TCR repertoire which they use as a pattern recognition receptor. This V-gene pairing in conjunction with the limited junctional diversity enables Vγ9/Vδ2 TCR to recognize families of unprocessed antigens with conserved molecular patterns. This pattern recognition by the Vγ9/Vδ2 TCR allows cytokine secretion by a large number of memory γ/δ cells in response to a diverse array of bacterial species. In addition these memory T cells are capable of responding within 2 h to alkylamine and organophosphate antigens produced by microbes. Similarly, the TCR repertoire of mucosal Vδ1 T cells is highly restricted, and it is likely that these cells recognize conserved structures which are expressed by stressed cells or activated dendritic cells. Thus, it is attractive to suppose that both Vγ9/Vδ2 and mucosal Vδ1 T cells function as a bridge linking innate and adaptive immune responses [3]. Furthermore, the new insights into the functions and activation pathways of γ/δ T cells have led to several therapeutic human trials in the fight against malignancies.

Vγ9/Vδ2 T Cells

Vγ9/Vδ2 T Cells Are Expanded by Various Microbes

Vγ9/Vδ2 T cells are the dominant γ/δ T cell population in the circulation and usually comprise 2–5% of human peripheral blood T cells. It has to be kept in mind that there are two nomenclatures for Vγ chains; according to an alternative nomenclature, the Vγ9 chain is also termed Vγ2 [4]. It was shown that Vγ9/Vδ2 T cells expand up to 50-fold and even higher during microbial infections. The highest expansions are found during infectious diseases, e.g. tuberculosis, salmonellosis, leishmaniasis, erlichiosis, brucellosis, tularemia, listeriosis, toxoplasmosis and malaria [5]. In addition Vγ9/Vδ2 T cells were shown to be stimulated by the herpes simplex virus (HSV) [6]. Furthermore, γ/δ T cells recovered from the genital lesions of patients with HSV infection expressed predominantly Vγ9/Vδ2, suggesting that circulating γ/δ T cells migrate to the genital lesions [7]. In contrast, Vγ9/Vδ2 T cells are not elevated in other viral diseases like CMV or HIV. In these diseases Vδ1 T cells predominate (see below). Thus, Vγ9/Vδ2 T cells are likely to play an important role in host defense of distinct microbial infections.

Vγ9/Vδ2 T Cells Are Activated by Non-Peptide Antigens

Already in 1990 it was shown, that Vγ9Vδ2 T cells can be activated by low molecular-weight, phosphate-containing non-peptide molecules [8]. Further studies demonstrated that three distinct groups of molecules are potent Vγ9Vδ2 activators: (1) Alkylphosphates like isopentenyl pyrophosphate (IPP) and related prenyl pyrophosphate derivatives, which are intermediates of the mevalonate pathway [9]; (2) Alkylamines which do not contain phosphate compounds and are ubiquitously present in plants, bacteria and also eukaryotes, and (3) Synthetic aminobisphosphonates (n-BPs) which are already in clinical use as therapeutic agents in osteoporosis and bone metastasis.

This response to non-peptidic antigens has so far only been observed in humans, rhesus macaques (*Macaca mulatta*) and night monkeys (*Aotus nancymaae*) [10–12]. In contrast to primate γ/δ T cells, murine γ/δ T cells do not respond to nonpeptide alkylamine or organophosphate antigens. Therefore, it is not at all certain that the biological functions of mouse γ/δ T cells are the same as those in primates [13].

The recognition of these agents was shown to be TCR mediated and did not require antigen-processing and presentation by conventional MHC molecules as opposed to the conventional recognition of small antigenic peptides by α/β T cells [1]. Notably these non-peptide antigens are shared by microbial and mammalian cells, raising the question why Vγ9/Vδ2 T cells are not constantly proliferating to high numbers in vivo and how the immune system can differentiate between friend and foe. It was suggested that normal levels of non-peptide antigens are too low to trigger proliferation but are perhaps high enough to keep these cells in a state of quiescent readiness. Only the induced overexpression of non-peptide antigens by infected cells or tumor cells would lead to the activation of γ/δ T cells [14]. However, recent evidence suggests that intermediates of the non-mevalonate pathway, which are only produced by a large array of microbes and not by mammalian cells (see below) [15–19], are much more potent Vγ9Vδ2 activators than IPP or alkylamines.

Non-Mevalonate Pathway Intermediates Are the Most Potent Vγ9/Vδ2 Activators

Until recently, IPP was thought to be synthesized exclusively via the mevalonate pathway (fig. 1). Now we know of the existence of a second isoprenoid pathway. This pathway is called the non-mevalonate (or Rohmer) pathway and later has been named after its key intermediate, the 2-C-methyl-d-erythritol 4-phosphate (MEP) pathway. In eukaryotes, archaebacteria, and certain eubacteria, biosynthesis of IPP proceeds entirely via the classical mevalonate pathway, while in many eubacteria and the plastids of algae and higher plants, IPP is supplied by the MEP pathway [16]. Recently, intermediates of the second isoprenoid biosynthesis pathway, i.e. 3-formyl-1-butyl pyrophosphate [19] and ((E)-4-hydroxy-3-methyl-but-2-enyl diphosphate) (HMB-PP) were identified as the strongest Vγ9/Vδ2 T cell activators [17, 20–22] which are several orders of magnitude more potent than IPP. This raised the question if such intermediates and not IPP are the major activators for Vγ9/Vδ2 T cells under physiologic conditions [23].

Bacterial species that lack the MEP pathway and synthesize IPP via the classical mevalonate pathway, such as *Streptococcus*, *Staphylococcus* and *Borrelia*, are unable to produce HMB-PP and do not specifically activate Vγ9/Vδ2 T cells. In contrast, *Listeria monocytogenes* which is the only pathogenic bacterium known to contain both pathways, activates Vγ9/Vδ2 T cells [20]. Disrupting the MEP pathway resulted in a loss of Vγ9/Vδ2 bioactivity whereas the disruption of the mevalonate pathway results in an increased activation (accumulation of HMB-PP). The confirmation that HMB-PP and not other intermediates of the MEP pathway are responsible for the specific activation of Vγ9/Vδ2 T cells was shown by studies with genetically engineered *Escherichia coli* strains, with essential enzymes of the MEP pathway being disrupted or deleted from the genome [15]. None of the known intermediates of the MEP pathway like DOXP, MEP, CDP-ME, CDP-MEP, and MEcPP had any T cell stimulatory activity. Thus, HMB-PP is the major activator for Vγ9Vδ2 cells within the MEP pathway.

Vγ9/Vδ2 T cells also expand in protozoal diseases like malaria, leishmaniasis, cryptosporidiosis or toxoplasmosis. In the case of malaria it was demonstrated that *Plasmodium falciparum* harbors the MEP pathway in a specialized subcellular organelle called apicoplast, which is evolutionarily derived from the chloroplast [23]. Thus, in analogy to bacterial infection, the production of HMB-PP in the apicoplasts of parasites might also be responsible for the stimulation of Vγ9Vδ2 T cells. In conclusion, it is possible that under physiologic conditions HMB-PP is the major activator for Vγ9/Vδ2 T cells and represents a 'danger' signal for the immune system after microbial infection. This would allow a

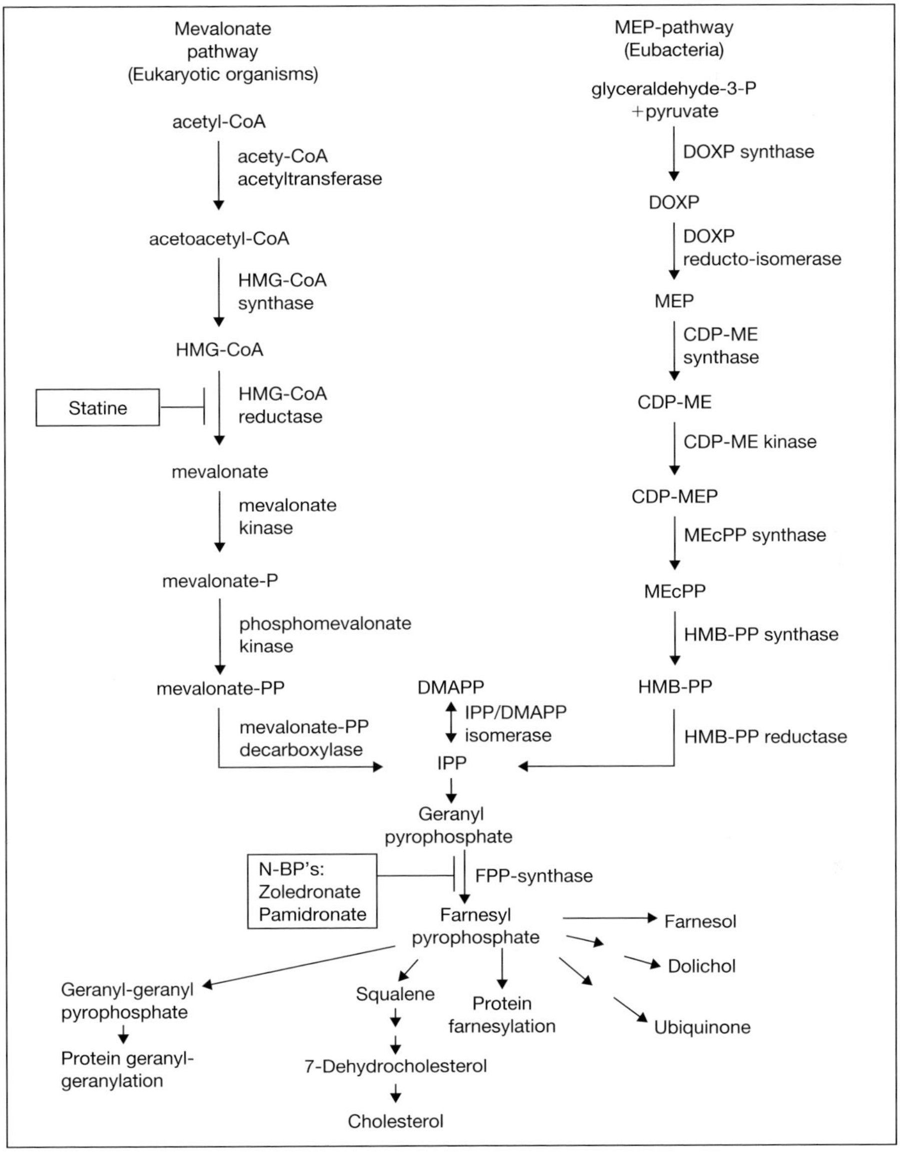

Fig. 1. Isoprenoid biosynthesis in *L. monocytogenes* via the classical mevalonate pathway (left) and the alternative MEP pathway (right). CDP-ME = 4-diphosphocytidyl-2-C-methyl-*D*-erythritol; CDP-MEP = 4-diphosphocytidyl-2-C-methyl-*D*-erythritol 2-phosphate; DMAPP = dimethylallyl pyrophosphate; DOXP = 1-deoxy-*D*-xylulose 5-phosphate; FPP = farnesyl diphosphate; HMB-PP = (E)-4-hydroxy-3-methyl-but-2-enyl-pyrophosphate; HMG-CoA = 3-hydroxy-3-methylglutaryl-coenzyme A; IPP = isopentenyl pyrophosphate; LMW = low molecular weight; n-BPs = nitrogen-containing bisphosphonates; MEP = 2-C-methyl-*D*-erythritol 4-phosphate; MEcPP = 2-C-methyl-*D*-erythritol 2,4-cyclopyrophosphate; P = phosphate.

specific but broad reactivity towards many microbial pathogens, by targeting a vital metabolic route shared by most human pathogens. Activation through IPP is less effective than HMB-PP but might be important in situations where there is a significant accumulation caused by treatment with bisphosphonates (see below) or overexpression of IPP after malignant transformation of a cell.

Alkylamines Activate Vγ9/Vδ2 T Cells

Alkylamines are the second group of molecules which are capable to stimulate Vγ9/Vδ2 T cells [24, 25]. Interestingly, they do not contain any phosphates and are ubiquitously found in the blood, urine, breast milk, vaginal secretions, and amniotic fluid of healthy individuals. They are secreted by commensal and pathogenic bacteria but are also found at high concentrations in tea and at lower concentrations in other edible plant products such as mushrooms, apples, and wine. One such alkylamine, ethylamine, is found in brewed tea and in its precursor form, *L*-theanine. It was suggested that tea drinking primes γ/δ T cells to mediate memory cytokine responses to antigens shared by tea and bacteria [24]. However, unlike HMB-PP/IPP or n-BP's, tea drinking does not lead to a proliferation of γ/δ T cells in PBMCs. Instead, ingestion of tea was shown to boost the capacity of peripheral blood γ/δ T cells to secrete IFN-γ after in vitro restimulation with this alkylamine antigen [24]. These data provide evidence that dietary intake of tea and perhaps other vegetables and fruits containing alkylamine antigens may prime human γ/δ T cells to provide natural resistance to microbial infections and perhaps tumors.

N-Bisphosphonates Stimulate Vγ9/Vδ2 T Cells

In addition to prenyl pyrophosphate antigen, it was clearly demonstrated that nitrogen-containing bisphosphonates (n-BPs) are potent in vivo and in vitro stimulators of Vγ9/Vδ2 T cells [26, 27]. However, there is now strong evidence that n-BPs have a different mode of action than IPP or alkylamines and do not directly stimulate the Vγ9/Vδ2 TCR by mimicking IPP or HMB-PP as previously suggested [27, 28]. Instead, it is more likely that the activation is indirect by the inhibition of the farnesyl diphosphate (FPP) synthase which leads to an accumulation of intracellular IPP (fig. 1) [14, 29]. FPP synthase is an enzyme in the mevalonate pathway which is required for the synthesis of cholesterol and the isoprenoid lipids FPP and geranylgeranyl diphosphate (GGPP). By inhibiting FPP synthase, n-BPs prevent the post-translational modification (isoprenylation) of small GTP-binding proteins that are necessary for osteoclast function. The effectiveness of different n-BPs for eliciting a response by γ/δ T cells matched the order of potency for inhibiting human FPP synthase [29, 30]. Furthermore, the stimulatory effect of n-BPs on γ/δ T cell proliferation was consistently abrogated by simultaneous treatment with mevastatin, which inhibits the mevalonate pathway upstream of FPP synthase (HMG-CoA reductase) [14, 29]. In contrast to IPP and alkylamines which need to be continuously present in the media, n-BPs can be pulsed onto antigen presenting cells and still retain their ability to activate γ/δ T cells after the cells has been washed [14, 31]. Thus, n-BPs require internalization in order to exert their effects, further supporting the hypothesis that n-BPs do not directly stimulate Vγ9/Vδ2 T cells and mimic alkylphosphates or IPP.

Vγ9/Vδ2 T Cells Can Kill Bacteria within Hours after Activation

Th1 type γ/δ T cells produce IFN-γ and TNF-α very rapidly upon ligand recognition which provide an important stimulus for macrophages at the early phase of infection [32, 33]. Thus, effector functions of activated γ/δ T cells already take place before they start proliferating. By using a chimeric severe combined immunodeficiency (SCID) mouse (hu-SCID) model, it was shown that human Vγ9/Vδ2 T cells mediate resistance to extracellular gram-positive (*Staphylococcus aureus*) and gram-negative (*E. coli* and *Morganella morganii*) bacteria [34]. This resistance was evident already 1 day after infection, and bacteria were cleared well before γ/δ T cell expansion was detected 6 days after infection, further confirming that expansion of γ/δ T cells is not required for an antibacterial response. In addition, intravenous treatment of infected, reconstituted hu-SCID mice with the n-BP pamidronate markedly increased the in vivo antibacterial effect of Vγ9/Vδ2 T cells. The data further support the notion that Vγ9/Vδ2 T cells are important mediators of natural immunity against extracellular bacterial infection and may bridge the gap between innate and acquired immunity.

The Vγ9/Vδ2 TCR Repertoire Is Shaped by Non-Peptide Antigens

It is still open for discussion if non-peptide antigens directly bind to the Vγ9/Vδ2 TCR and if a presenting element exists. Studies showed that there is clearly a requirement for a species-specific cell-cell contact [35]. However, none of the known antigen-presenting elements like MHC class I and II or CD1 were required for T cell activation suggesting that a novel presenting element might exist. Strong support for a direct recognition of non-peptide antigens with the Vγ9Vδ2 TCR comes from studies which demonstrated that a transfected Vγ9Vδ2 TCR can confer responsiveness on a hitherto unresponsive cell [36]. Furthermore, antibodies to the γ/δ TCR blocked recognition [37]. In addition, the CDR3 regions of Vγ9/Vδ2 chains were critical in determining reactivity to non-peptide antigens since changes of the amino acid sequence within the junctional region abolished recognition of phosphoantigens [38]. Vγ9/Vδ2 TCRs of the vast majority of γ/δ T cells in adult blood exhibit shared features in their CDR3 regions: preferential usage of Jγ1.2 segment with short N-regions [28, 39] and a conserved hydrophobic residue (L, V or I) at position 97 encoded by the VDJ junction [40].

Cord blood cells stimulated with pyrophosphomonoester antigen in vitro showed preferential expansion of γ/δ T cells expressing Vγ9/Vδ2 TCR chains compared with those stimulated with the polyclonal mitogen phytohemagglutinin [39]. Since the Vγ9/Vδ2 T cells predominate in the γ/δ T cell population of adults, but not in cord blood, it was postulated that the repertoire of adult γ/δ T cells might be selected by environmental antigens after birth [41, 42]. Such a developmental repertoire selection by non-peptide antigens may enable Vγ9/Vδ2 T cells to respond to microbial infection rapidly in a manner resembling innate immunity. This hypothesis is in line with our data that the circulating Vδ2 TCR repertoire of healthy adults shows multiple oligoclonal expansions which are stable over time [43]. However, the final proof that non-peptide antigens and the Vγ9/Vδ2 chain co-crystallize is still missing [44]. Furthermore, it was demonstrated that a single Vγ9Vδ2 T cell clone could recognize alkylamines and IPP which are differently charged molecules [24].

Vγ9/Vδ2 T Cells and Tumor Surveillance

An important aspect of antigen specificity of human γ/δ cells is their capacity to recognize and kill tumor targets. T cells expressing the Vγ9/Vδ2 heterodimer recognize bone marrow-derived tumor cells such as the non-Hodgkin B cell lymphoma line Daudi both in vitro and in a SCID animal model in vivo [1, 45]. Vγ9/Vδ2 T cells also recognize and kill the B cell lymphoma RPMI-8233, the T cell lymphoma MOLT-4 and the erythroleukemia line K562 [14].

Different hematological malignancies and mammary carcinoma cells were reported to have an increased expression and function of HMG-CoA reductase, the rate limiting enzyme in the mevalonate pathway. It was suggested that recognition of cells which overproduce nonpeptidic phosphorylated metabolites generated by the mevalonate pathway may allow the immune system to target cells with significant metabolic abnormalities [14]. Mevalonate intermediates are necessary for sterol synthesis, cell growth, and membrane integrity, and thus tumor cells which up-regulate this pathway become more visible to the immune system. This would explain a general mechanism whereby one T cell population could survey all tissues for both normal and transformed cells that are metabolically altered in the mevalonate pathway.

There have been several reports, that infection with mycobacteria in patients bearing malignancy may modify the course of tumor progression. In this regard application of the bacillus of Calmette and Guerin (BCG) was thought to have a role in the treatment of several cancers in the late 1950s and 1960s [46]. However, controlled studies failed to confirm these early observations and modern chemotherapy and radiotherapy led to the abandonment of the use of BCG for most cancers. Nevertheless, BCG still has an established role in the treatment of bladder cancer and 70% response rates were shown for carcinomas in situ [46]. The mechanism of action is not known; however, it was suggested that non-peptide antigens from mycobacteria might stimulate γ/δ T cells which in turn inhibit tumor growth [47, 48]. It is known that infection with BCG or *Mycobacterium tuberculosis* cause a rapid expansion of Vγ9/Vδ2 T cells in the peripheral blood of humans and non-human primates [12, 49].

Activation of Vγ9/Vδ2 T Cells as a Therapeutic Approach in Tumor Treatment: From Bench to Bedside

In 1996, Lamb et al. [50] observed that those leukemia patients which survived bone marrow transplantation had much higher numbers of γ/δ T cells in their peripheral blood than those which had a relapse of their disease. It was suggested that γ/δ T cells improved disease-free survival following transplantation. Three years later Kunzmann et al. [51] reported that bisphosphonates, which are widely used to inhibit osteoclastic bone resorption, can stimulate the Vγ9/Vδ2 T cells in the peripheral blood. Treatment with bisphosphonates resulted in a substantial increase in the percentage of γ/δ T cells in the peripheral blood (up to 70%) and was accompanied with a secretion of IFN-γ. Pamidronate-activated γ/δ T cells exhibited in vitro specific cytotoxicity against lymphoma (Daudi) and myeloma cell lines (RPMI, U266) [26]. In addition, pamidronate treated bone marrow cultures of patients with multiple myeloma showed significantly reduced plasma cell survival compared with untreated cultures, suggesting that γ/δ T cells, which are activated by bisphosphonates might contribute to the anti-tumor effects in patients with lymphoid malignancies. To evaluate this anti-tumor activity of γ/δ T cells in vivo, the same group initiated a pilot study of low dose IL-2 in combination with pamidronate in patients with refractory low grade non-Hodgkin lymphoma or multiple lymphoma [52]. Three of 9 patients who showed a vigorous in vitro proliferation of γ/δ T cells in response to pamidronate/IL-2 achieved objective responses. Administration of pamidronate and low dose IL-2 was well tolerated. Notably, 9 other patients did not exhibit any in vitro or in vivo stimulation of γ/δ T cells, indicating that the function of γ/δ T cells can be impaired in some patients with lymphoid malignancies. In conclusion, γ/δ T cell-mediated immunotherapy is feasible and can induce objective tumor responses in a subset of patients.

The potential anti-tumor effects of activated Vγ9/Vδ2 T cells have led to pilot studies in patients with solid tumors as well. An Italian group initiated a phase I study of the bisphosphonate zoledronic acid in patients with metastatic breast cancer and prostate cancer [53]. In vitro data on neuroblastoma [54], bladder cancer [48] and human colon carcinoma cell lines suggest that γ/δ T cells might also have the capacity to efficiently kill other solid tumors as well [54]. Our own results have demonstrated in vivo anti-tumor activity of Vγ9/Vδ2 T cells against melanoma and pancreatic adenocarcinoma upon adoptive transfer into SCID mice transplanted with human tumors [55]. Interestingly, heat treated mistletoe extracts which are often used in palliative cancer treatment were shown to activate and to induce proliferation of Vγ9/Vδ2 T cells in vitro. This activity was sensitive to treatment with alkaline phosphatase but not with proteinase K, indicating that the ligands must be non-proteinaceous phosphate containing compounds [56, 57]. The hope to find a new way to treat malignancies by stimulating Vγ9/Vδ2 T cells even led to the foundation of a French company called 'Innate Pharma' (www.innate-pharma.com). They developed a synthetic analog of phosphorylated non-mevalonate pathway intermediates termed bromohydrin pyrophosphate [58] (Phosphostim™). This substance is currently in a phase I clinical trial in metastatic renal cell carcinoma. In addition, this company also uses an ex vivo approach to generate larger numbers of activated γ/δ T cells which are infused back to the patient. No data about the success of this approach are available at this point. Taken together, there is substantial hope that the in vivo activation of γδ T cells with phosphoantigen or aminobisphosphonate in combination with adoptive transfer of ex vivo expanded γδ T cells will lead to new therapeutic options for certain tumors [59].

Vδ1 Cells

Vδ1 T Cells Are the Dominant γ/δ T Cell Population at Mucosal Surfaces

γ/δ T cells which express the Vδ1 chain associated with various Vγ elements predominate at mucosal surfaces like the intestine and are located within the epithelial layer of the small and large intestine [60]. Vδ1 T cells are a minor population in the peripheral blood. In contrast to peripheral Vγ9/Vδ2 T cells there is no unique Vγ chain which is preferentially paired with Vδ1, and Vδ1 T cells can not be activated by non-peptide antigens like IPP, alkylamines or nBPs [31, 37, 40, 61]. Based on their anatomical localization it has been suggested that Vδ1 γ/δ T cells are 'watchdogs' of the intestinal tract and crucial players involved in immune surveillance. As pathogens cross the physical barriers of the host, both the innate and adaptive immune system are activated and Vδ1 γ/δ T cells are thought to link both immune responses. The activating antigens which bind to the Vδ1 TCR remain largely unknown. Recent evidence suggests, however, that these cells recognize stress-induced self antigens, which would enable many γ/δ T cells, carrying identical T cell receptors (TCR), to monitor multiple insults to the epithelium [62–65]. This is in line with our own observations that γ/δ T cells of healthy adults are clonally expanded within mucosal surfaces [43, 60, 66–68] and the skin [69].

Vδ1 T Cells Are Activated by MICA/B

MICA and MICB are distant relatives of MHC class I molecules which have no role in antigen presentation [63]. These are stress induced antigens expressed under the control of heat shock responsive promoters by intestinal epithelia and epithelial tumors [70]. They are thought to function as signals of cellular distress since they are increasingly expressed in infected and transformed epithelial cells (see below) [71]. In healthy individuals, expression of MIC is restricted to the intestinal epithelium [70]. It remains unclear, however, whether this tissue distribution is the result of permanent cellular stress or whether it may be induced by environmental factors. MICA/B are highly polymorphic molecules and, up to now, there are at least 54 recognized human MICA alleles [72].

The natural receptor of MICA/B is NKG2D which is expressed on natural killer (NK) cells and CD8+ α/β T cells but also on a subset of γ/δ T cells [73]. IL-15 which is secreted by stressed epithelial cells seems to be a key cytokine in the induction of NKG2D [74, 75] and the activation of γ/δ T cells [76]. Intestinal γ/δ T cells which express the Vδ1 chain have been shown to recognize the stress inducible MICA/B molecule [62, 77]. Interestingly, also peripheral γ/δ T cells expressing the Vδ2 chain were shown to be activated by MICA after infection with *M. tuberculosis* [71]. A recent paper suggested that the γ/δ TCR itself directly binds to MICA/B [78]. Thus, MICA/B delivers both the TCR-dependent signal 1 and the NKG2D-dependent costimulatory signal 2 for a subset of γ/δ T cells. These dual receptor interactions of MIC might serve to preclude erroneous T cell activation by cross-reactive cell surface determinants [78]. However, this hypothesis is still controversial.

Vδ1 T Cells Are Activated by Glycolipids Presented by CD1

During host infection, microbial antigens such as LPS provide critical innate signals for dendritic cell (DC) maturation *via* cell surface Toll-like receptors (TLRs) [79]. However, maturation of DCs by microbial stimuli alone may promote a short lived burst of IL-12 leading to an exhausted DC population unable to produce this cytokine upon subsequent encounter with naive T cells in the lymph node [80]. The ability of the host to successfully defend against microbial invasion is, therefore, dependent upon both innate signals provided by microbial products as well as T cell signals [81]. Recent data demonstrate that a subpopulation of human Vδ1 γ/δ T cells can selectively stimulate dendritic cells to undergo maturation [80, 82, 83]. These Vδ1 γ/δ T cells specifically recognize CD1c on the surface of immature DC or B cells [84]. Upon recognition of CD1c, they secrete TNF-α and other proinflammatory factors that, together with microbial products such as LPS, induce immature DC to mature and to produce IL-12 [80, 82]. Furthermore, TNF-α and IFN-γ, secreted by activated γ/δ T cells, provide the critical signals for Th1 polarization of naive CD4+ α/β T cells. In addition, CD1c specific γ/δ T cells are cytolytic in nature using both perforin and Fas-mediated cytotoxicities [84]. CD1 molecules are known to present lipid and glycolipid complexes to T cells [85–87]. Importantly recognition of CD1c occurred in the absence of exogenous foreign antigens [81, 84], implying reactivity against yet to be identified self-lipids [87]. Recognition of glycolipids presented by CD1 is not limited to γ/δ T cells. It was reported that NKT cells expressing Vα24 in humans and Vα14 in mice can also be activated by glycolipids [87]. Thus, γ/δ T cells are involved during the early stages of an immune response by interacting with immature DC and act as a link between innate (DC) and acquired (α/β T cells) immunity.

Lipid Extracts from Gram-Negative Bacteria Indirectly Stimulate Vδ1 T Cells

Another recent in vitro study demonstrated that peripheral Vδ1 γ/δ T cells can be activated and expanded in response to lipid extracts of gram-negative bacteria in the presence of monocyte-derived dendritic cells [88]. Whereas Vδ1 T cells could be expanded up to 64 fold, this was not the case for Vδ2 cells. The bioactive molecules could not be determined but there was evidence that LPS might be the activating agent. The direct dendritic cell contact was essential to mediate Vδ1 T cell activation, but the major antigen presenting molecules like MHC or CD1 did not play a role. It was suggested that lipid extracts of gram-negative bacteria, like LPS, do not directly stimulate γ/δ T cells. Instead LPS activates dendritic cells through TLRs resulting in a marked secretion of IL-12 which provides a major cytokine signal for driving Vδ1 T cell proliferation. However, the proliferation of Vδ1 T cells was not unspecific but rather highly specific since proliferation could be selectively blocked by Vδ1 specific mAb. mAb against MICA or NKG2D only showed partial inhibition suggesting that activated dendritic cells might express other, unknown costimulatory molecules which are specifically recognized by the Vδ1 TCR. Thus, there must be several distinct antigens which are capable of activating Vδ1 T cells.

TCR Repertoire of Vδ1 T Cells

In contrast to the peripheral blood Vδ1 is the dominant V region expressed by γ/δ T cells in human intestine [89] whereas only a minority express Vδ2 or Vδ3. In addition, several Vα genes can occasionally recombine with Cδ [43]. Sequencing of rearranged junctional regions from Vδ1 transcripts have shown that the TCR repertoire of intestinal γ/δ T cells, which were derived from healthy adults, is highly restricted [43, 66]. The method which we have used to analyze the TCR repertoire of intestinal γ/δ T cells was the CDR3 length analysis (CDR3 spectratyping; fig. 2). Identical dominant γ/δ T cell clones could be identified along the entire colon. Moreover, almost identical fingerprint like CDR3 profiles were obtained from different parts of the colon suggesting a model in which γ/δ T cell clones, selected by ligands in the colon undergo expansion and recirculation before lodging throughout the entire colon (fig. 3) [67, 90]. γ/δ T cells are not only compartmentalized based on their V region usage but also on their rearranged junctional regions. There was no overlap between the Vδ1 TCR repertoire of circulating γ/δ T cells and that of intestinal Vδ1 T cells [66]. Different dominant Vδ1 T cell clones were present in both compartments, suggesting that distinct antigens are recognized and that Vδ1 T cells from the peripheral blood are not just recirculating γ/δ T cells homing back to the intestine.

Similar to Vγ9/Vδ2 T cells of the circulation also Vδ1 T cells from the intestine are highly polyclonal at birth and showed increasing restriction with age [67]. By the age of 14–17, the repertoire was oligoclonal and resembled the repertoire of individuals in the sixth to seventh decade. This was also the case for Vδ2 T cells. The recognized antigens are still not known but several potential candidates have been discussed above. CDR3 spectratyping of Vδ2 transcripts from the colon and peripheral blood of healthy adults demonstrated that oligoclonal expansions were present in both compartment but similar to the Vδ1 repertoire there was no overlap (fig. 3) [43, 90]. Thus, Vδ2 T cells of the colon are not just in equilibrium with circulating Vγ9/Vδ2 T cells which

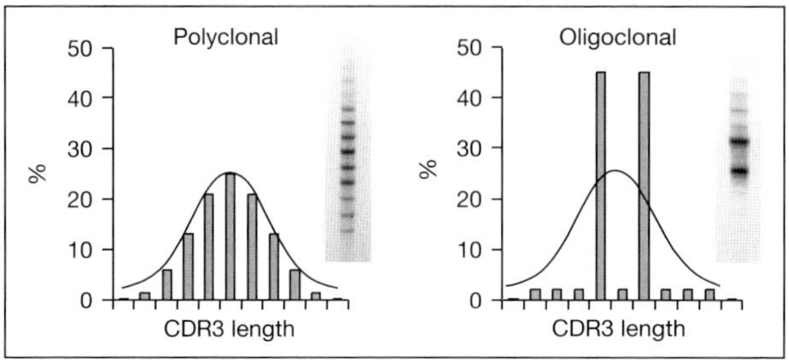

Fig. 2. Based on the broad range and frequency distribution of TCR δ CDR3 lengths, the band pattern of a polyclonal TCR δ repertoire has a bell-curved distribution in which the greatest band intensity is at the median length, with bands containing shorter and longer CDR3 regions being approximately equally distributed in decreasing frequency on either side (left). Adjacent bands were shown to be separated by a distance corresponding to 3 bp (i.e. one codon). An oligoclonal CDR3 profile is characterized by dominant bands of different lengths which do not follow a bell curve distribution.

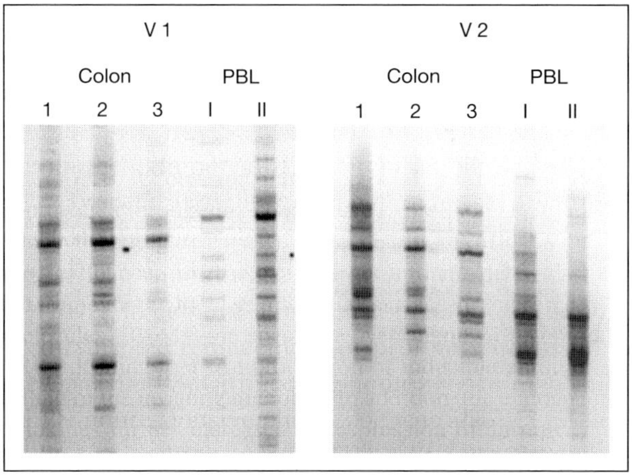

Fig. 3. CDR3 length analysis of TCR Vδ1 and Vδ2 transcripts in human colon (1 = cecum, 2 = transverse colon, 3 = sigmoid) and peripheral blood lymphocytes (PBL sample I and II taken 1 min apart) from a representative healthy subject. Each of the three colon sites analyzed for Vδ1 or Vδ2 yielded an almost identical CDR3 profile with multiple dominant bands. The CDR3 profile of the colon was clearly distinct from that in the PBL. Sequence analysis confirmed that there was no overlap of identical TCR δ transcripts between both sites.

migrate to the intestine. Furthermore, the Vδ1 and Vδ2 TCR repertoires from the intestine and the circulation were stable over a period of at least 18 months [43]. In analogy to intestinal Vδ1 T cells dominant Vδ2 T cell clones showed a homogenous, fingerprint like distribution along the entire colon, suggesting that also this minor population must be expanded by local antigens and recirculates continuously [67]. A local expansion of γ/δ T cells without recirculation would not result in a homogenous distribution but rather in a patchy distribution of different dominant clones along the large intestine. In conclusion, intestinal and circulating γ/δ T cells, expressing identical V regions, are highly distinct populations and recognize distinct antigens. It is likely that recirculating intestinal Vδ1 and Vδ2 populations were not detected in the peripheral blood because their numbers were below the detection limit of the method used.

In order to gain more insight into the compartmentalization of γ/δ T cells within the organism we chose the pig as an animal model which allows more extensive experiments [68, 91, 92]. We could confirm the human data that γ/δ T cells are polyclonal at birth and show increasing restriction with age irrespective of the V region analyzed. In addition, we were able to analyze the γ/δ TCR repertoires from the stomach, duodenum, Ileum, Peyer's patches, jejunum, mesenteric lymph nodes, lungs, spleen and thymus. Different TCR δ repertoires were present between the lungs and the intestinal mucosa but also within different parts of the gastrointestinal tract. These data clearly indicated that γ/δ T cells within different compartments of the intestinal tract recognize distinct antigens and might have distinct functions as well. If all intestinal γ/δ T cells would recognize the same antigens throughout the intestine, there would be no compartmentalization. Probably γ/δ T cells interact with different self antigens which are specific for that site or with foreign antigens which are limited to that site. For example, the microbial colonization of the colon is distinct from that of the small intestine. Interestingly, we occasionally observed identical TCR δ transcripts in the intestine and the lungs. These γ/δ T cells are likely to transport immunological information between different compartments of the immune system and might protect the lung against microbial infections of the intestine [93].

The Role of Vδ1 T Cells in Microbial Infections

Although Vδ1 T cells are a minor population within the peripheral blood they can be activated and expanded by infectious diseases. When renal allograft recipients developed cytomegalovirus (CMV) infection, Vδ1 and Vδ3 T cell subpopulations markedly expanded within the peripheral blood [94].

Furthermore, Vδ1 and Vδ3 T cells from CMV-infected subjects were able to proliferate in vitro in the presence of CMV. This expansion was inhibited by anti-γ/δ TCR mAbs. Viral glycoproteins are thought to be the recognized antigens. Infection with HIV-1 also leads to a marked proliferation of Vδ1 γ/δ T cells in the peripheral blood [95, 96]. At the same time a decrease in Vγ9/Vδ2 T cells was observed and was accompanied by a selective anergy of these cells in response to non-peptide antigens [97]. It was suggested that the activation and expansion of circulating Vδ1 γ/δ T cells during the course of CMV and HIV infection might primarily occur in the intestinal epithelia followed by the migration of these activated cells to the peripheral blood [98]. This hypothesis is supported by studies with non-human primates which could demonstrate that infection with simian immunodeficiency virus by the rectal mucosal route was followed by a significant increase in γ/δ T cells [99]. Alternatively it was hypothesized that B cells from the peripheral blood of HIV+ subjects specifically activate Vδ1 T cells [100]. Hence, it is not known if the increased numbers of Vδ1 T cells in the peripheral blood are caused by the expansion of circulating Vδ1 T cells or by the migration and influx of activated mucosal Vδ1 T cells.

It has been reported that Vδ1 expressing γ/δ T cells present in the synovial fluid of patients with Lyme arthritis proliferate in response to lysates of *B. burgdorferi* [101–103].The ligands recognized by Vδ1 T cells appear to be lipidated hexapeptides derived from Borrelia outer surface proteins [103]. Stimulation was not restricted by class I or class II MHC, CD1a, CD1b or CD1c. Interestingly no selective expansion of Vδ1 T cells was observed within the Borrelia stimulated PBMC [101]. This data further support the notion that Vδ1 T cells from the peripheral blood are distinct from the Vδ1 subpopulations in different tissues. Similarly Vδ1 T cells predominate the dermal granulomas in patients with American cutaneous leishmaniasis [104], indicating that there are antigens which specifically expand Vδ1 and not Vδ2 expressing γ/δ T cells.

The Role of Vδ1 T Cells in Tumor Recognition

T lymphocytes bearing the Vδ1 T cell receptor can infiltrate solid cancers and exhibit a selective lytic activity against a variety of tumor cell lines like colorectal cancer, esophageal cancer, renal cell cancer, pancreatic cancer and lung cancer [77, 105–109]. The mode of action is unknown, but it is thought that the γ/δ T cells recognize self antigens that are overexpressed at the tumor site. It is not known if these Vδ1 T cells are resident, local T cells which are activated by transformed cells or if they are derived from circulating Vδ1 T cells which are attracted by chemokines. Studies with autologous melanoma xenografts in SCID mice demonstrated that Vδ1 cells might be more important for the anti-tumor effect of certain tumors than Vδ2 cells [110], although we have observed anti-melanoma activity of Vδ2 T cells in a similar SCID model [55]. After intravenous infusion of lymphoid cells only Vδ1 T cells were found at the tumor site and inhibited the growth of human melanoma. Vδ2 T cells were only detectable in the spleen of SCID mice [110]. The apparent discrepancy with our own results [55] indicates that the contribution of Vδ1 vs. Vδ2 T cells in anti-tumor activity may vary even in the case of a given tumor entity.

A prime candidate for a stress-induced self antigen which is overexpressed by tumor cells is MICA. It was shown that MICA is recognized by tumor infiltrating T cells [77, 111]. Human NK cells and Vδ1 T cells bearing NKG2D receptors can lyse tumor cells which frequently express MICA [73, 112, 113]. Furthermore, immobilized MICA could induce the proliferation of human ovarian or colonic carcinoma-derived Vδ1 cells which were able to kill tumor cells [113]. This proliferation could be inhibited by anti-MICA mAb. Thus, MICA selectively triggers the expansion of Vδ1 T cells. The cytotoxic effects on tumor cells are known to be enhanced by IFN-γ and γ/δ T cells were shown to be an early source of IFN-γ, which in turn upregulates the production of IFN-γ by α/β T cells [114]. Furthermore, mice lacking γ/δ cells are highly susceptible to multiple regimens of cutaneous carcinogenesis [115, 116]. After exposure to carcinogens, skin cells express Rae-1 and H60, major histocompatibility complex-related molecules structurally resembling human MICA. In vitro, skin-associated NKG2D+ γ/δ cells killed skin carcinoma cells by a mechanism that was sensitive to blocking NKG2D engagement. Several cancers seem to evade this immune response by producing large amounts of soluble MICA which binds to NKG2D. As a consequence, NKG2D is downregulated and the responsiveness of tumor-antigen specific effector γ/δ T cells is severely impaired [117]. Anti-tumor effects of Vδ1 T cells might not be limited to solid tumors. In vitro data suggest that they are also capable of lysing acute lymphoblastic leukemia (ALL) [118, 119] and Burkitt's lymphoma cells [120, 121]. The activating antigens are not known. Thus, Vδ1 cells may be also useful in cellular immunotherapy against leukemia.

Migration and Homing of γ/δ T Cells

Current models suggest that α/β T cells which selectively recirculate through an organ will preferentially express adhesion receptors for organ specific endothelial ligands [122]. For example, skin homing T cells express the cutaneous lymphocyte antigen (CLA) which binds to E-selectin, which is expressed on endothelial cells of the skin. T cells homing to the gut express α4β7 integrin which binds to MAdCAM-1 (mucosal addressin cell adhesion molecule-1) [123]. Similar to α/β T cells also γ/δ T cells are targeted to different organs guided by a spatiotemporal sequence of tissue homing receptors, local induction and/or selection processes [124, 125]. Selection may be exerted by local antigens, cytokines, chemokines and cell-matrix interactions. This leads to the expansion and maintenance of some clones, whereas others are diluted out or deleted. The spatial compartmentalization of γ/δ T cells in different microenvironments has major functional consequences and leads to a partial fragmentation of immunoregulatory circuits at the local level.

During the last years it became evident that chemokines play an essential role in the fine tuning of migration and homing. Both circulating and resident γ/δ T cells seem to express a distinct pattern of chemokine receptors, which allow them to reach distinct sites in response to specific stimuli. As explained below in more detail γ/δ T cells themselves can produce a number of chemokines, thereby contributing possibly to the recruitment of circulating lymphocytes. Furthermore, the production of chemokines by endothelial cells might be induced by cytokines released by activated γ/δ T cells. It was shown that Vδ1 and Vδ2 T cells secrete a wide variety of cytokines, in particular IFN-γ. Thus, this would provide an amplification loop in the recruitment of leukocytes during inflammation.

Chemokine Expression of Peripheral γ/δ T Cells

Peripheral γ/δ T cells have the capacity just like peripheral α/β T cells to migrate in response to chemokines indicating that they are recruited during inflammation depending on their expressed chemokine receptors [4, 126]. Chemokines are likely to function in concert with antigen-TCR binding to promote migration specificity and subsequent lymphocyte expansion and/or retention of γ/δ T cells at inflammatory sites. In contrast to peripheral α/β T cells, the majority of Vγ9/Vδ2 T cells strongly express CCR5, mainly in conjunction with the memory marker CD45RO [4, 127]. In addition, CXCR3 is expressed on most peripheral γ/δ T cells. One of the ligands for CXCR3 is IP-10 which may effectively contribute to the γ/δ migratory pattern [127–129]. Interestingly, IP-10 is highly expressed in the tuberculosis-affected lung and the lymph node. Thus, IP-10 may attract CXCR3 positive Vγ9/Vδ2 T cells which contribute to the containment of *Mycobacterium tuberculosis* infection. In response to specific recognition of mycobacterial antigens by the Vγ9/Vδ2 TCR, γ/δ T cells secrete cytokines like INF-γ which are important in granuloma formation [130, 131].

Importantly, the cell surface expression of chemokine receptors is modulated by cytokines and antigen recognition via the TCR [127, 132–135]. It was shown that CXCR3 and CCR5 which are highly expressed on unstimulated peripheral Vγ9Vδ2 T cells, were down-modulated after stimulation with the non-peptide antigen IPP or by TCR triggering [127, 132, 133, 135, 136]. In contrast CCR7 is upregulated [135]. In lymph nodes Vδ2 T cells have a naïve phenotype (CD45RA+) and express CCR7 but not CCR5 or CXCR3 [136]. Thus, it was suggested that the stimulation of memory Vγ9/Vδ2 T cells induces a lymph-node homing program. Interestingly Vδ2 T cells from inflammatory fluids express to a large proportion the naïve phenotype CD45RA+, CCR5+ and CXCR3+ but not CCR7+ [136]. Thus, it was suggested that memory CD45RO+ Vγ9Vδ2 T cells can differentiate into CD45RA+ cells. However, so far no CD45RA+ γ/δ T cells were recovered from IPP-stimulated cultures of CD45RO+ cells [136]. Stimulation of peripheral Vγ9/Vδ2 T cells was also shown to induce a prominent expression of essential B cell costimulatory molecules and to provide potent B cell help during in vitro antibody production [135]. Thus, γ/δ T cells have a role in humoral immunity during microbial infections which is in part induced by changes in the chemokine pattern.

α/β T cells expressing CD45RO can be divided into effector memory cells which combat spread of the pathogen and memory T cells which guard against subsequent infections ('central memory'). It was demonstrated that memory T cells express CCR7 whereas memory effector cells do not [137, 138]. The $CCR7^-$ cells produce effector cytokines such as INF-γ, IL4 and IL-5. It was shown that peripheral γ/δ T cells expressing the Vγ9/Vδ2 TCR can also be differentiated towards distinct memory phenotypes depending on the activating cytokine [139]. Vγ9/Vδ2 T cells stimulated in the presence of IL-2 or IL-15 produced abundant amounts of the pro-inflammatory cytokines TNF-α and IFN-γ. In addition, they expressed CCR5 but not CCR7. In contrast, Vγ9/Vδ2 T cells stimulated in the presence of IL-21 did not produce TNF-α or IFN-γ but expressed the chemokine receptor CCR7. Thus, memory γ/δ T cells do not constitute one homogenous population but rather are comprised of subsets that may be characterized by differential expression of additional markers like chemokine receptors.

γ/δ T Cells Can Be Polarized into TH1/TH2 Cells

In α/β T cells expression of certain chemokine receptors has been preferentially associated with a TH1 (CXCR3,CCR5) or a TH2 (CCR3,CCR4) skewed phenotype [140–142]. Similarly, circulating γ/δ T cells can be polarized into

TH1 or TH2 cells [143, 144] with distinct patterns of chemokine expression. TH2 producing γ/δ T cells had lower levels of CXCR3 and higher levels of CCR4 when compared to TH1 γ/δ T cells [129]. However, this is in contrast to another study which could not detect a change in chemokine receptor expression after polarization of Vδ2/Vγ9 T cells into TH1 or TH2 subsets [145].

Chemokine Expression of Mucosal γ/δ T Cells

As mentioned above, the major adhesion molecule for the homing of intestinal T cells is the α4β7 integrin [146]. However, this does not explain the observed compartmentalization of T cells within the gastrointestinal tract [68]. The expression of particular sets of chemokine receptors provides leucocytes with an exclusive combinatorial address code for positioning within the tissue in a multistep navigation mode. The chemokine CCR9 was shown to be expressed on murine γ/δ and α/β T cells as well as human α/β T cells of the small intestine and the thymus, whereas colonic T cells lack CCR9 [147]. The natural ligand of CCR9 is CCL25 which is expressed by epithelial cells of the small intestine and the thymus [148–151].

Currently there are no published studies on the expression of CCR9 on human γ/δ T cells, but our own preliminary data indicate expression on activated circulating Vδ1 but not Vδ2 T cells [Marischen et al., unpubl. obs.]. However, some data are available from murine studies. For example, CCR9 k.o. mice have increased numbers of peripheral γ/δ T cells but reduced numbers of intraepithelial γ/δ and CD8+ α/β T cells of the small intestine [147, 152]. This reduction of small intestinal γ/δ T cells was accompanied with a preferential loss of Vγ5/Vδ4 expressing γ/δ T cells indicating that the expression of CCR9 was limited to subsets of γ/δ T cells which expressed certain TCR [147, 153]. Anti-CCL25 antibody-treated mice also showed a 50% reduction of CD8+ γ/δ and α/β T cells [154]. Thus, CCR9/CCL25 is important in the recruitment of γ/δ T cells to the small intestinal mucosa.

In addition, CCR9/CCL25 might be important in the formation of cryptopatches and the consequent appearance of IEL [155]. Recent data suggest that Peyer's patch dendritic cells imprint gut homing specificity on α/β T cells by inducing the expression of α4β7 and CCR9 [156]. It remains to be established whether this is also the case for γ/δ T cells. The expressed TCR also seems to play a role in the homing of γ/δ T cells. Antibody-mediated TCR stimulation enhanced CCL25 responsiveness, indicating that CCL25 induced migration of γ/δ T cells is augmented by TCR signaling [153]. In addition, the expression of CD69 further increased the migration of T cells in response to

CCL25. These findings are consistent with previous results demonstrating that the activation state of T cells can influence their response to chemokines.

The role of additional chemokines/ligands in directing T cells to other sites of the intestinal tract is much less clear [157]. Recent evidence suggests that CCL28 which is expressed by epithelial cells of the colon, lungs and the oral cavity attracts IgA producing cells expressing CCR10 [158, 159]. However, only few T cells in any mucosal tissue examined express CCR10 [160]. Thus, it is likely that T cells which home for example to the colon express other, yet undefined chemokine receptors. In contrast, the chemokines responsible for recruiting T cells to the skin are better known. Skin homing T cells express the chemokine receptors CCR10 or CCR4 [161, 162] which bind to CCL27 or CCL17 respectively [162, 163]. The selective expression of chemokines at distinct sites may permit functional specialization of immune responses.

γ/δ T Cells Can Have Immunosuppressive and Anti-Inflammatory Activities

There is good evidence that mucosal γ/δ T cells play a key role in the induction of oral tolerance [164, 165]. Murine in vitro studies showed that γ/δ T cells can suppress cytotoxic T lymphocyte (CTL) responses to diverse antigens. The inhibitory activity of γ/δ T cells is very potent, since a very small number of γ/δ T cells can transfer mucosal tolerance to naïve mice [166] and less than 10 cells per culture inhibit over 90% of the CTL response [165]. This inhibitory response by γ/δ T cells was shown to be caused by soluble mediators and not by cell-cell contact. The mediators are not identified but IL-10 and TGF-β were suggested to be potential candidates. In the decidua γ/δ T cells increase in number during pregnancy and were shown to be mainly Vδ1 [167]. The recognized antigen might be presented by HLA-G [168]. During early pregnancy these γ/δ T cells expressed high levels of IL-10 and TGF-β, suggesting that γ/δ T cells in normal pregnancy create a milieu promoting immunotolerance to the fetus by defending the trophoblast from cytotoxic effector mechanisms [169]. Similarly γ/δ T cells are thought to prevent airway hyper-responsiveness in patients with asthma by inhibiting the inflammatory response [170–173]. This is supported by studies which could demonstrate that human γ/δ T cells from bronchoalveolar lavage secrete fibroblast growth factor 9 (FGF-9) and keratinocyte growth factor (KGF) [174]. Thus, γ/δ T cells are part of the protective regulation of the airways and play a role in the epithelial tissue reconstitution of the lung.

Analogous to the γ/δ T cells of the airways, intraepithelial γ/δ T cells of the intestine were also suggested to down modulate inflammatory responses and to function in repair of damaged epithelial tissues by secreting epithelial growth factors [83, 175–179]. Furthermore, several recent papers support the protective or regulatory role of γ/δ T cells in murine models of inflammatory diseases which lack γ/δ T cells [172, 177, 180–182]. γ/δ k.o. or γ/δ T cell-depleted mice showed much stronger inflammation and died earlier than wild type mice. Finally, the downmodulation of the inflammatory α/β T cell response to bacterial infection [183] and the killing of activated macrophages were described [184]. In an in vitro culture model, we observed inhibitory activity of human phosphoantigen-stimulated γ/δ T cells on the proliferative response of CD4+ α/β T cells to tetanus toxoid or bacterial superantigens [Wesch et al., unpubl. results]. Thus, subsets of γ/δ T cells are likely to have an important role in the suppression and termination of an ongoing immune response.

Interestingly, under certain conditions these regulatory γ/δ T cells might have deleterious effects, e.g. by preventing the rejection of a tumor [164, 185, 186]. For example, γ/δ T cells were shown to attenuate the activity of CTL in a mouse tumor model whereas depletion of γ/δ T cells resulted in a regression of the tumor. These regulatory γ/δ T cells were shown to transcribe IL-10 and TGF-β. Blocking with mAb confirmed that the inhibitory action of γ/δ T cells on CTL was at least partly mediated by IL-10 and TGF-β. These data are in line with others which demonstrated that regulatory γ/δ T cells can inhibit tumor rejection whereas depletion of γ/δ T cells augmented the ability of the mice to reject the tumor [187]. The elimination of the tumor was further improved in IL-10 knockout mice suggesting that IL10-secreted by γ/δ T cells might suppress tumor rejection. The inhibition of CTL might also prevent the clearance of fungal infections. In mice infected with *Cryptococcus neoformans* [188], the number of live microorganisms was significantly reduced in the lungs of mice depleted of γ/δ T cells or γ/δ knockout mice. Thus, different subsets of γ/δ T cells can have completely distinct functions like killing or immunomodulatory effects.

Expression of Toll-Like Receptors

Innate immune cells express TLRs as pattern recognition receptors to respond to pathogen-associated molecular pattern such as LPS, peptidoglycan, flagellin, single-stranded RNA or bacterial DNA [189]. Although not typically expressed on cells of the adaptive immune system, several reports have shown that T cells can express certain TLRs. $CD25^{high}$ regulatory T cells (Treg) were thus found to express TLR4 and to directly respond to the TLR4 ligand LPS

[190]. Moreover, murine γ/δ T cells proliferated in response to lipid A, due to the expression of functional TLR2 [191]. We have screened a number of established human Vδ1 and Vδ2 γ/δ T cell clones for the expression of TLRs by RT-PCR. Apart from TLR1 which was ubiquitously expressed, we observed a strong signal for TLR3, the receptor for double-stranded viral DNA [192]. Using a TLR3-specific mAb we also detected TLR3 protein intracellulary as well as small amounts on the cell surface, as shown by flow cytometry. Functional studies using polyinosinic-polycytidylic acid (poly I:C) as a surrogate TLR3 ligand revealed costimulatory effects in response to phosphoantigens and survival-enhancing effects also on highly purified γ/δ T cell lines [Wesch et al., unpubl. obs.]. This is seemingly at variance with a recent study by Kunzmann et al. [193], where the stimulatory effects of poly I:C on human γ/δ T cells were described as indirect effects, due to the type I IFN released by small numbers of contaminating dendritic cells. The γ/δ lines analyzed in this study, however, did not express TLR3. In view of the suspected role of γ/δ T cells as a link between innate and adaptive immune system, it seems important to clarify the expression and potential role of the various TLRs in subpopulations of γ/δ T cells, which is under way in our laboratory.

Concluding Remarks

Despite tremendous progress over the years, the significance of γ/δ T cells in health and disease is still not completely clear. The emerging picture suggests that these cells fulfill important functions at the interface between innate and adaptive immune responses. At one hand, γ/δ T cells express antigen receptors made up on the basis of rearranging genes. In contrast to α/β T cells; however, this TCR is not used to recognize myriads of peptides in an MHC-restricted fashion, but rather is used as a pattern recognition receptor to recognize (and rapidly respond to) pathogen-associated molecular patterns such as bacterial phosphoantigens, or stress-induced self antigens such as MICA/B. Data from both the murine and the human systems indicate that γ/δ T cells located at mucosal surfaces are there to control the integrity of surrounding epithelial tissue. Preferential localization of one or the other γ/δ T cell subset seems to be guided by the expression of integrins and chemokine receptors. In the future, the interactions between local γ/δ T cells and epithelia should be investigated in more detail in order to better understand the physiological role of γ/δ T cells. Furthermore, the potential application of γ/δ T cells for the immunotherapy of certain types of cancer will be investigated in order to explore new therapeutic modalities. Encouraging results from experimental and clinical studies indicate that a detailed understanding of the molecular mechanisms how γ/δ T cells recognize epithelial cells and control epithelial integrity might eventually lead to γ/δ T cell-based immunotherapies.

References

1 Hayday AC: γ/δ cells: A right time and a right place for a conserved third way of protection. Annu Rev Immunol 2000;18:975–1026.
2 Cai JL, Tucker PW: γ/δ T cells: Immunoregulatory functions and immunoprotection. Chem Immunol 2001;79:99–138.
3 Mak TW, Ferrick DA: The γ/δ T-cell bridge: Linking innate and acquired immunity. Nat Med 1998;4:764–765.
4 Kabelitz D, Wesch D: Features and functions of γδ T lymphocytes: Focus on chemokines and their receptors. Crit Rev Immunol 2003;23:339–370.
5 Kabelitz D, Wesch D, Hinz T: γ/δ T cells, their T cell receptor usage and role in human diseases. Springer Semin Immunopathol 1999;21:55–75.
6 Bukowski JF, Morita CT, Brenner MB: Recognition and destruction of virus-infected cells by human γ/δ CTL. J Immunol 1994;153:5133–5140.
7 Verjans GM, Roest RW, Van Der KA, Van Dijk G, Van Der Meijden WI, Osterhaus AM: Isopentenyl pyrophosphate-reactive Vγ9Vδ2 T helper 1-like cells are the major γ/δ T cell subset recovered from lesions of patients with genital herpes. J Infect Dis 2004;190:489–493.
8 Pfeffer K, Schoel B, Plesnila N, Lipford GB, Kromer S, Deusch K, Wagner H: A lectin-binding, protease-resistant mycobacterial ligand specifically activates Vγ9+ human γ/δ T cells. J Immunol 1992;148:575–583.
9 Morita CT, Lee HK, Leslie DS, Tanaka Y, Bukowski JF, Marker-Hermann E: Recognition of nonpeptide prenyl pyrophosphate antigens by human γ/δ T cells. Microbes Infect 1999;1: 175–186.
10 Chen ZW, Letvin NL: Adaptive immune response of Vγ2Vδ2 T cells: A new paradigm. Trends Immunol 2003;24:213–219.
11 Wang H, Lee HK, Bukowski JF, Li H, Mariuzza RA, Chen ZW, Nam KH, Morita CT: Conservation of nonpeptide antigen recognition by rhesus monkey Vγ2Vδ2 T cells. J Immunol 2003;170:3696–3706.
12 Shen Y, Zhou DJ, Qiu LY, Lai XM, Simon M, Shen L, Kou ZC, Wang QF, Jiang JM, Estep J, Hunt R, Clagett M, Sehgal PK, Li YY, Zeng XJ, Morita CT, Brenner MB, Letvin NL, Chen ZW: Adaptive immune response of Vγ2Vδ2(+) T cells during mycobacterial infections. Science 2002;295: 2255–2258.
13 Mestas J, Hughes CC: Of mice and not men: Differences between mouse and human immunology. J Immunol 2004;172:2731–2738.
14 Gober HJ, Kistowska M, Angman L, Jeno P, Mori L, De Libero G: Human T cell receptor γδ cells recognize endogenous mevalonate metabolites in tumor cells. J Exp Med 2003;197:163–168.
15 Altincicek B, Moll J, Campos N, Foerster G, Beck E, Hoeffler JF, Grosdemange-Billiard C, Rodriguez-Concepcion M, Rohmer M, Boronat A, Eberl M, Jomaa H: Cutting edge: Human γδ T cells are activated by intermediates of the 2-C-methyl-*D*-erythritol 4-phosphate pathway of isoprenoid biosynthesis. J Immunol 2001;166:3655–3658.
16 Eberl M, Hintz M, Reichenberg A, Kollas AK, Wiesner J, Jomaa H: Microbial isoprenoid biosynthesis and human γδ T cell activation. FEBS Lett 2003;544:4–10.
17 Eberl M, Jomaa H, Hayday AC: Integrated immune responses to infection – Cross-talk between human γδ T cells and dendritic cells. Immunology 2004;112:364–368.
18 Feurle J, Espinosa E, Eckstein S, Pont F, Kunzmann V, Fournie JJ, Herderich M, Wilhelm M: *Escherichia coli* produces phosphoantigens activating human γ/δ T cells. J Biol Chem 2002;277: 148–154.
19 Belmant C, Espinosa E, Poupot R, Peyrat MA, Guiraud M, Poquet Y, Bonneville M, Fournie JJ: 3-Formyl-1-butyl pyrophosphate A novel mycobacterial metabolite-activating human γδ T cells. J Biol Chem 1999;274:32079–32084.

20 Begley M, Gahan CG, Kollas AK, Hintz M, Hill C, Jomaa H, Eberl M: The interplay between classical and alternative isoprenoid biosynthesis controls γδ T cell bioactivity of *Listeria monocytogenes*. FEBS Lett 2004;561:99–104.

21 Altincicek B, Duin EC, Reichenberg A, Hedderich R, Kollas AK, Hintz M, Wagner S, Wiesner J, Beck E, Jomaa H: LytB protein catalyzes the terminal step of the 2-C-methyl-*D*-erythritol-4-phosphate pathway of isoprenoid biosynthesis. FEBS Lett 2002;532:437–440.

22 Kollas AK, Duin EC, Eberl M, Altincicek B, Hintz M, Reichenberg A, Henschker D, Henne A, Steinbrecher I, Ostrovsky DN, Hedderich R, Beck E, Jomaa H, Wiesner J: Functional characterization of GcpE, an essential enzyme of the non-mevalonate pathway of isoprenoid biosynthesis. FEBS Lett 2002;532:432–436.

23 Eberl M, Jomaa H: A genetic basis for human γδ T-cell reactivity towards microbial pathogens. Trends Immunol 2003;24:407–409.

24 Bukowski JF, Morita CT, Brenner MB: Human γδ T cells recognize alkylamines derived from microbes, edible plants, and tea: Implications for innate immunity. Immunity 1999;11:57–65.

25 Kamath AB, Wang L, Das H, Li L, Reinhold VN, Bukowski JF: Antigens in tea-beverage prime human Vγ2Vδ2 T cells in vitro and in vivo for memory and nonmemory antibacterial cytokine responses. Proc Natl Acad Sci USA 2003;100:6009–6014.

26 Kunzmann V, Bauer E, Feurle J, Weissinger F, Tony HP, Wilhelm M: Stimulation of γδ T cells by aminobisphosphonates and induction of antiplasma cell activity in multiple myeloma. Blood 2000;96:384–392.

27 Das H, Wang L, Kamath A, Bukowski JF: Vγ2Vδ2 T-cell receptor-mediated recognition of aminobisphosphonates. Blood 2001;98:1616–1618.

28 Miyagawa F, Tanaka Y, Yamashita S, Mikami B, Danno K, Uehara M, Minato N: Essential contribution of germline-encoded lysine residues in Jγ1.2 segment to the recognition of nonpeptide antigens by human γδ T cells. J Immunol 2001;167:6773–6779.

29 Thompson K, Rogers MJ: Statins prevent bisphosphonate-induced γδ-T-cell proliferation and activation in vitro. J Bone Miner Res 2004;19:278–288.

30 Sanders JM, Ghosh S, Chan JM, Meints G, Wang H, Raker AM, Song Y, Colantino A, Burzynska A, Kafarski P, Morita CT, Oldfield E: Quantitative structure-activity relationships for γδ T cell activation by bisphosphonates. J Med Chem 2004;47:375–384.

31 Green AE, Lissina A, Hutchinson SL, Hewitt RE, Temple B, James D, Boulter JM, Price DA, Sewell AK: Recognition of nonpeptide antigens by human Vγ 9Vδ2 T cells requires contact with cells of human origin. Clin Exp Immunol 2004;136:472–482.

32 Lang F, Peyrat MA, Constant P, Davodeau F, David-Ameline J, Poquet Y, Vie H, Fournie JJ, Bonneville M: Early activation of human Vγ9Vδ2 T cell broad cytotoxicity and TNF production by nonpeptidic mycobacterial ligands. J Immunol 1995;154:5986–5994.

33 Wang L, Das H, Kamath A, Bukowski JF: Human Vγ2Vδ2 T cells produce IFN-γ and TNF-α with an on/off/on cycling pattern in response to live bacterial products. J Immunol 2001;167:6195–6201.

34 Wang L, Kamath A, Das H, Li L, Bukowski JF: Antibacterial effect of human Vγ2Vδ2 T cells in vivo. J Clin Invest 2001;108:1349–1357.

35 Kato Y, Tanaka Y, Tanaka H, Yamashita S, Minato N: Requirement of species-specific interactions for the activation of human γδ T cells by pamidronate. J Immunol 2003;170:3608–3613.

36 Tanaka Y, Morita CT, Nieves E, Brenner MB, Bloom BR: Natural and synthetic non-peptide antigens recognized by human γδ T cells. Nature 1995;375:155–158.

37 Tanaka Y, Sano S, Nieves E, De Libero G, Rosa D, Modlin RL, Brenner MB, Bloom BR, Morita CT: Nonpeptide ligands for human γδ T cells. Proc Natl Acad Sci USA 1994;91:8175–8179.

38 Bukowski JF, Morita CT, Band H, Brenner MB: Crucial role of TCR γ chain junctional region in prenyl pyrophosphate antigen recognition by γδ T cells. J Immunol 1998;161:286–293.

39 Yamashita S, Tanaka Y, Harazaki M, Mikami B, Minato N: Recognition mechanism of nonpeptide antigens by human γδ T cells. Int Immunol 2003;15:1301–1307.

40 Nishimura H, Hirokawa M, Fujishima N, Fujishima M, Miura I, Sawada K: Contribution of complementarity-determining region 3 of the T-cell receptor Vδ2 chain to the recognition of aminobisphosphonates by human γδ T-cells. Int J Hematol 2004;79:369–376.

41 Parker CM, Groh V, Band H, Porcelli SA, Morita C, Fabbi M, Glass D, Strominger JL, Brenner MB: Evidence for extrathymic changes in the T cell receptor gamma/delta repertoire. J Exp Med 1990; 171:1597–1612.

42 De Rosa SC, Andrus JP, Perfetto SP, Mantovani JJ, Herzenberg LA, Herzenberg LA, Roederer M: Ontogeny of γδ T cells in humans. J Immunol 2004;172:1637–1645.

43 Holtmeier W, Chowers Y, Lumeng A, Morzycka-Wroblewska E, Kagnoff MF: The δ T cell receptor repertoire in human colon and peripheral blood is oligoclonal irrespective of V region usage. J Clin Invest 1995;96:1108–1117.

44 Allison TJ, Garboczi DN: Structure of γδ T cell receptors and their recognition of non-peptide antigens. Mol Immunol 2002;38:1051–1061.

45 Fisch P, Malkovsky M, Kovats S, Sturm E, Braakman E, Klein BS, Voss SD, Morrissey LW, DeMars R, Welch WJ: Recognition by human Vγ9/Vδ2 T cells of a GroEL homolog on Daudi Burkitt's lymphoma cells. Science 1990;250:1269–1273.

46 Alexandroff AB, Jackson AM, O'Donnell MA, James K: BCG immunotherapy of bladder cancer: 20 years on. Lancet 1999;353:1689–1694.

47 Alexandroff AB, Black J, Bollina P, Esuvaranathan K, James K: Are γδ T lymphocytes involved in intravesical bacillus Calmette-Guerin immunotherapy of bladder cancer? Biochem Soc Trans 1997;25:363S.

48 Kato Y, Tanaka Y, Miyagawa F, Yamashita S, Minato N: Targeting of tumor cells for human γδ T cells by nonpeptide antigens. J Immunol 2001;167:5092–5098.

49 Kabelitz D, Bender A, Prospero T, Wesselborg S, Janssen O, Pechhold K: The primary response of human γ/δ+ T cells to *Mycobacterium tuberculosis* is restricted to Vγ9-bearing cells. J Exp Med 1991;173:1331–1338.

50 Lamb LS Jr, Henslee-Downey PJ, Parrish RS, Godder K, Thompson J, Lee C, Gee AP: Increased frequency of TCR γδ+ T cells in disease-free survivors following T cell-depleted, partially mismatched, related donor bone marrow transplantation for leukemia. J Hematother 1996;5: 503–509.

51 Kunzmann V, Bauer E, Wilhelm M: γ/δ T-cell stimulation by pamidronate. N Engl J Med 1999;340:737–738.

52 Wilhelm M, Kunzmann V, Eckstein S, Reimer P, Weissinger F, Ruediger T, Tony HP: γδ T cells for immune therapy of patients with lymphoid malignancies. Blood 2003;102:200–206.

53 Dieli F, Gebbia N, Poccia F, Caccamo N, Montesano C, Fulfaro F, Arcara C, Valerio MR, Meraviglia S, Di Sano C, Sireci G, Salerno A: Induction of γ/δ T-lymphocyte effector functions by bisphosphonate zoledronic acid in cancer patients in vivo. Blood 2003;102: 2310–2311.

54 Schilbach K, Geiselhart A, Handgretinger R: Induction of proliferation and augmented cytotoxicity of γδ T lymphocytes by bisphosphonate clodronate. Blood 2001;97:2917–2918.

55 Kabelitz D: Characterization of tumor reactivity of human Vγ9Vδ2 γδ T-cells in vitro and in severe combined immunodeficiency mice in vivo. J Immunol 2004;173:6767–6776.

56 Fischer S, Scheffler A, Kabelitz D: Activation of human γ/δ T-cells by heat-treated mistletoe plant extracts. Immunol Lett 1996;52:69–72.

57 Fischer S, Scheffler A, Kabelitz D: Stimulation of the specific immune system by mistletoe extracts. Anticancer Drugs 1997;8(suppl 1):S33–S37.

58 Espinosa E, Belmant C, Pont F, Luciani B, Poupot R, Romagne F, Brailly H, Bonneville M, Fournie JJ: Chemical synthesis and biological activity of bromohydrin pyrophosphate, a potent stimulator of human γδ T cells. J Biol Chem 2001;276:18337–18344.

59 Kabelitz D, Wesch D, Pitters EZM: Potential of human γδ T-lymphocytes for immmuntherapy of cancer. Int J Cancer 2004;112:727–732.

60 Holtmeier W: Compartmentalization of γ/δ T cells and their putative role in mucosal immunity. Crit Rev Immunol 2003;23:473–488.

61 Bukowski JF, Morita CT, Tanaka Y, Bloom BR, Brenner MB, Band H: Vγ2Vδ2 TCR-dependent recognition of non-peptide antigens and Daudi cells analyzed by TCR gene transfer. J Immunol 1995;154:998–1006.

62 Groh V, Steinle A, Bauer S, Spies T: Recognition of stress-induced MHC molecules by intestinal epithelial γ/δ T cells. Science 1998;279:1737–1740.

63 Spies T: Induction of T cell alertness by bacterial colonization of intestinal epithelium. Proc Natl Acad Sci USA 2002;99:2584–2586.
64 Pardoll DM: Immunology – Stress, NK receptors, and immune surveillance. Science 2001;294: 534–536.
65 Havran WL, Chien YH, Allison JP: Recognition of self antigens by skin-derived T cells with invariant γ/δ antigen receptors. Science 1991;252:1430–1432.
66 Chowers Y, Holtmeier W, Harwood J, Morzycka-Wroblewska E, Kagnoff MF: The Vδ1 T cell receptor repertoire in human small intestine and colon. J Exp Med 1994;180:183–190.
67 Holtmeier W, Witthöft T, Hennemann A, Winter HS, Kagnoff MF: The TCR-δ repertoire in human intestine undergoes characteristic changes during fetal to adult development. J Immunol 1997; 158:5632–5641.
68 Holtmeier W, Käller J, Geisel W, Pabst R, Caspary WF, Rothkötter HJ: Development and compartmentalization of the porcine TCR δ repertoire at mucosal and extraintestinal sites: The pig as a model for analyzing the effects of age and microbial factors. J Immunol 2002;169: 1993–2002.
69 Holtmeier W, Pfänder M, Hennemann A, Zollner TM, Kaufmann R, Caspary WF: The TCR δ repertoire in normal human skin is restricted and distinct from the TCR δ repertoire in the peripheral blood. J Invest Dermatol 2001;116:275–280.
70 Groh V, Bahram S, Bauer S, Herman A, Beauchamp M, Spies T: Cell stress-regulated human major histocompatibility complex class I gene expressed in gastrointestinal epithelium. Proc Natl Acad Sci USA 1996;93:12445–12450.
71 Das H, Groh V, Kuijl C, Sugita M, Morita CT, Spies T, Bukowski JF: MICA engagement by human Vγ2Vδ2 T cells enhances their antigen-dependent effector function. Immunity 2001;15:83–93.
72 Steinle A, Groh V, Spies T: Diversification, expression, and γδ T cell recognition of evolutionarily distant members of the MIC family of major histocompatibility complex class I-related molecules. Proc Natl Acad Sci USA 1998;95:12510–12515.
73 Bauer S, Groh V, Wu J, Steinle A, Phillips JH, Lanier LL, Spies T: Activation of NK cells and T cells by NKG2D, a receptor for stress-inducible MICA. Science 1999;285:727–729.
74 Roberts AI, Lee L, Schwarz E, Groh V, Spies T, Ebert EC, Jabri B: NKG2D receptors induced by IL-15 costimulate CD28-negative effector CTL in the tissue microenvironment. J Immunol 2001;167:5527–5530.
75 Ohteki T: Critical role for IL-15 in innate immunity. Curr Mol Med 2002;2:371–380.
76 Garcia VE, Jullien D, Song M, Uyemura K, Shuai K, Morita CT, Modlin RL: IL-15 enhances the response of human γ/δ T cells to nonpeptide microbial antigens. J Immunol 1998;160:4322–4329.
77 Groh V, Rhinehart R, Secrist H, Bauer S, Grabstein KH, Spies T: Broad tumor-associated expression and recognition by tumor-derived γ/δ T cells of MICA and MICB. Proc Natl Acad Sci USA 1999;96:6879–6884.
78 Wu J, Groh V, Spies T: T cell antigen receptor engagement and specificity in the recognition of stress-inducible MHC class I-related chains by human epithelial γ/δ T cells. J Immunol 2002;169: 1236–1240.
79 Takeda K, Kaisho T, Akira S: Toll-like receptors. Annu Rev Immunol 2003;21:335–376.
80 Leslie DS, Vincent MS, Spada FM, Das H, Sugita M, Morita CT, Brenner MB: CD1-mediated γ/δ T cell maturation of dendritic cells. J Exp Med 2002;196:1575–1584.
81 Vincent MS, Leslie DS, Gumperz JE, Xiong X, Grant EP, Brenner MB: CD1-dependent dendritic cell instruction. Nat Immunol 2002;3:1163–1168.
82 Ismaili J, Olislagers V, Poupot R, Fournie JJ, Goldman M: Human γ/δ T cells induce dendritic cell maturation. Clin Immunol 2002;103:296–302.
83 Jameson J, Witherden D, Havran WL: T-cell effector mechanisms: γ/δ and CD1d-restricted subsets. Curr Opin Immunol 2003;15:349–353.
84 Spada FM, Grant EP, Peters PJ, Sugita M, Melian A, Leslie DS, Lee HK, van Donselaar E, Hanson DA, Krensky AM, Majdic O, Porcelli SA, Morita CT, Brenner MB: Self-recognition of CD1 by γ/δ T cells: Implications for innate immunity. J Exp Med 2000;191:937–948.
85 Beckman EM, Porcelli SA, Morita CT, Behar SM, Furlong ST, Brenner MB: Recognition of a lipid antigen by CD1-restricted αaβ+ T cells. Nature 1994;372:691–694.

86 Matsuda JL, Kronenberg M: Presentation of self and microbial lipids by CD1 molecules. Curr Opin Immunol 2001;13:19–25.
87 De Libero G, Mori L: Self glycosphingolipids: New antigens recognized by autoreactive T lymphocytes. News Physiol Sci 2003;18:71–76.
88 Das H, Sugita M, Brenner MB: Mechanisms of Vδ1 γδ T cell activation by microbial components. J Immunol 2004;172:6578–6586.
89 Deusch K, Luling F, Reich K, Classen M, Wagner H, Pfeffer K: A major fraction of human intraepithelial lymphocytes simultaneously expresses the γ/δ T cell receptor, the CD8 accessory molecule and preferentially uses the Vδ1 gene segment. Eur J Immunol 1991;21:1053–1059.
90 Holtmeier W, Hennemann A, May E, Duchmann R, Caspary WF: T cell receptor δ repertoire in inflamed and noninflamed colon of patients with IBD analyzed by CDR3 spectratyping. Am J Physiol 2002;282:G1024-G1034.
91 Thielke KH, Hoffmann-Moujahid A, Weisser C, Waldkirch E, Pabst R, Holtmeier W, Rothkötter HJ: Proliferating intestinal γ/δ T cells recirculate rapidly and are a major source of the γ/δ T cell pool in the peripheral blood. Eur J Immunol 2003;33:1649–1656.
92 Holtmeier W, Geisel W, Bernert K, Butler JE, Sinkora M, Rehakova Z, Sinkora J, Caspary WF: Prenatal development of the porcine TCR δ repertoire: Dominant expression of an invariant T cell receptor Vδ3-Jδ3 chain. Eur J Immunol 2004;34:1941–1949.
93 Dunkley M, Pabst R, Cripps A: An important role for intestinally derived T cells in respiratory defence. Immunol Today 1995;16:231–236.
94 Dechanet J, Merville P, Lim A, Retiere C, Pitard V, Lafarge X, Michelson S, Meric C, Hallet MM, Kourilsky P, Potaux L, Bonneville M, Moreau JF: Implication of γδ T cells in the human immune response to cytomegalovirus. J Clin Invest 1999;103:1437–1449.
95 Boullier S, Cochet M, Poccia F, Gougeon ML: CDR3-independent γδ Vδ1+ T cell expansion in the peripheral blood of HIV-infected persons. J Immunol 1995;154:1418–1431.
96 Hinz T, Wesch D, Friese K, Reckziegel A, Arden B, Kabelitz D: T cell receptor γδ repertoire in HIV-1-infected individuals. Eur J Immunol 1994;24:3044–3049.
97 Poccia F, Boullier S, Lecoeur H, Cochet M, Poquet Y, Colizzi V, Fournie JJ, Gougeon ML: Peripheral Vγ9/Vδ2 T cell deletion and anergy to nonpeptidic mycobacterial antigens in asymptomatic HIV-1-infected persons. J Immunol 1996;157:449–461.
98 Dechanet J, Merville P, Pitard V, Lafarge X, Moreau JF: Human γδ T cells and viruses. Microbes Infect 1999;1:213–217.
99 Lehner T, Mitchell E, Bergmeier L, Singh M, Spallek R, Cranage M, Hall G, Dennis M, Villinger F, Wang Y: The role of γ/δ T cells in generating antiviral factors and β-chemokines in protection against mucosal simian immunodeficiency virus infection. Eur J Immunol 2000;30:2245–2256.
100 Hyjek EM, Bartkowiak J, Drozdz R, Wasik TJ, Jasinski M, Kaneko Y, Lischner HW, Kozbor D: Evidence for B cell-mediated activation of Vδ1+ T lymphocytes during progression of HIV infection. J Immunol 1997;158:464–474.
101 Glatzel A, Entschladen F, Zollner TM, Kraiczy P, Brade V, Kaufmann R, Janssen O, Lengl-Janssen B, Wesch D, Kabelitz D: The responsiveness of human Vδ1 γδ T cells to *Borrelia burgdorferi* is largely restricted to synovial-fluid cells from patients with Lyme arthritis. J Infect Dis 2002;186: 1043–1046.
102 Vincent MS, Roessner K, Lynch D, Wilson D, Cooper SM, Tschopp J, Sigal LH, Budd RC: Apoptosis of Fashigh CD4+ synovial T cells by borrelia-reactive Fas-ligand(high) γδ T cells in Lyme arthritis. J Exp Med 1996;184:2109–2117.
103 Vincent MS, Roessner K, Sellati T, Huston CD, Sigal LH, Behar SM, Radolf JD, Budd RC: Lyme arthritis synovial γδ T cells respond to *Borrelia burgdorferi* lipoproteins and lipidated hexapeptides. J Immunol 1998;161:5762–5771.
104 Uyemura K, Deans RJ, Band H, Ohmen J, Panchamoorthy G, Morita CT, Rea TH, Modlin RL: Evidence for clonal selection of γ/δ T cells in response to a human pathogen. J Exp Med 1991; 174:683–692.
105 Ferrarini M, Heltai S, Chiesa G, Sabbadini MG: Vδ1+ γ/δ T lymphocytes infiltrating human lung cancer express the CD8 α/α homodimer. Scand J Immunol 1994;40:363–367.
106 Zocchi MR, Ferrarini M, Migone N, Casorati G: T-cell receptor Vδ gene usage by tumour reactive γδ T lymphocytes infiltrating human lung cancer. Immunology 1994;81:234–239.

107 Maeurer MJ, Martin D, Walter W, Liu K, Zitvogel L, Halusczcak KX, Rabinowich H, Storkus W, Lotze MT: Human intestinal Vδ1+ lymphocytes recognize tumor cells of epithelial origin. J Exp Med 1996;183:1681–1696.
108 Choudhary A, Davodeau F, Moreau A, Peyrat MA, Bonneville M, Jotereau F: Selective lysis of autologous tumor cells by recurrent γδ tumor-infiltrating lymphocytes from renal carcinoma. J Immunol 1995;154:3932–3940.
109 Thomas ML, Badwe RA, Deshpande RK, Samant UC, Chiplunkar SV: Role of adhesion molecules in recruitment of Vδ1 T cells from the peripheral blood to the tumor tissue of esophageal cancer patients. Cancer Immunol Immunother 2001;50:218–225.
110 Lozupone F, Pende D, Burgio VL, Castelli C, Spada M, Venditti M, Luciani F, Lugini L, Federici C, Ramoni C, Rivoltini L, Parmiani G, Belardelli F, Rivera P, Marcenaro S, Moretta L, Fais S: Effect of human natural killer and γδ T cells on the growth of human autologous melanoma xenografts in SCID mice. Cancer Res 2004;64:378–385.
111 Vetter CS, Groh V, Straten PT, Spies T, Brocker EB, Becker JC: Expression of stress-induced MHC class I related chain molecules on human melanoma. J Invest Dermatol 2002;118:600–605.
112 Ferrarini M, Ferrero E, Dagna L, Poggi A, Zocchi MR: Human γ/δ T cells: A nonredundant system in the immune-surveillance against cancer. Trends Immunol 2002;23:14–18.
113 Qi J, Zhang J, Zhang S, Cui L, He W: Immobilized MICA could expand human Vδ1 γδ T cells in vitro that displayed major histocompatibility complex class I chain-related A-dependent cytotoxicity to human epithelial carcinomas. Scand J Immunol 2003;58:211–220.
114 Gao Y, Yang W, Pan M, Scully E, Girardi M, Augenlicht LH, Craft J, Yin Z: γδ T cells provide an early source of interferon γ in tumor immunity. J Exp Med 2003;198:433–442.
115 Girardi M, Oppenheim DE, Steele CR, Lewis JM, Glusac E, Filler R, Hobby P, Sutton B, Tigelaar RE, Hayday AC: Regulation of cutaneous malignancy by γ/δ T cells. Science 2001;294: 605–609.
116 Girardi M, Glusac E, Filler RB, Roberts SJ, Propperova I, Lewis J, Tigelaar RE, Hayday AC: The distinct contributions of murine T cell receptor (TCR)γ/δ+ and TCRααβ+ T cells to different stages of chemically induced skin cancer. J Exp Med 2003;198:747–755.
117 Groh V, Wu J, Yee C, Spies T: Tumour-derived soluble MIC ligands impair expression of NKG2D and T-cell activation. Nature 2002;419:734–738.
118 Lamb LS Jr, Musk P, Ye Z, van Rhee F, Geier SS, Tong JJ, King KM, Henslee-Downey PJ: Human γδ(+) T lymphocytes have in vitro graft vs leukemia activity in the absence of an allogeneic response. Bone Marrow Transplant 2001;27:601–606.
119 Duval M, Yotnda P, Bensussan A, Oudhiri N, Guidal C, Rohrlich P, Boumsell L, Grandchamp B, Vilmer E: Potential antileukemic effect of γδ T cells in acute lymphoblastic leukemia. Leukemia 1995;9:863–868.
120 Hacker G, Kromer S, Falk M, Heeg K, Wagner H, Pfeffer K: Vδ1+ subset of human γδ T cells responds to ligands expressed by EBV-infected Burkitt lymphoma cells and transformed B lymphocytes. J Immunol 1992;149:3984–3989.
121 Fisch P, Oettel K, Fudim N, Surfus JE, Malkovsky M, Sondel PM: MHC-unrestricted cytotoxic and proliferative responses of two distinct human γ/δ T cell subsets to Daudi cells. J Immunol 1992;148:2315–2323.
122 Abitorabi MA, Mackay CR, Jerome EH, Osorio O, Butcher EC, Erle DJ: Differential expression of homing molecules on recirculating lymphocytes from sheep gut, peripheral, and lung lymph. J Immunol 1996;156:3111–3117.
123 von Andrian UH, Mackay CR: Advances in immunology: T-cell function and migration – Two sides of the same coin. N Engl J Med 2000;343:1020–1033.
124 Wilson E, Hedges JF, Butcher EC, Briskin M, Jutila MA: Bovine γ/δ T cell subsets express distinct patterns of chemokine responsiveness and adhesion molecules: A mechanism for tissue-specific γ/δ T cell subset accumulation. J Immunol 2002;169:4970–4975.
125 Chao CC, Sandor M, Dailey MO: Expression and regulation of adhesion molecules by γ/δ T cells from lymphoid tissues and intestinal epithelium. Eur J Immunol 1994;24:3180–3187.
126 Roth SJ, Diacovo TG, Brenner MB, Rosat JP, Buccola J, Morita CT, Springer TA: Transendothelial chemotaxis of human ααβ and γ/δ T lymphocytes to chemokines. Eur J Immunol 1998;28: 104–113.

127 Glatzel A, Wesch D, Schiemann F, Brandt E, Janssen O, Kabelitz D: Patterns of chemokine receptor expression on peripheral blood γ/δ T lymphocytes: Strong expression of CCR5 is a selective feature of Vδ2/Vγ9 γ/δ T cells. J Immunol 2002;168:4920–4929.
128 Romagnani P, Annunziato F, Lazzeri E, Cosmi L, Beltrame C, Lasagni L, Galli G, Francalanci M, Manetti R, Marra F, Vanini V, Maggi E, Romagnani S: Interferon-inducible protein 10, monokine induced by interferon γ, and interferon-inducible T-cell α-chemoattractant are produced by thymic epithelial cells and attract T-cell receptor (TCR) ααβ (+)CD8(+) single-positive T cells, TCR γ/δ (+) T cells, and natural killer-type cells in human thymus. Blood 2001;97:601–607.
129 Ferrero E, Biswas P, Vettoretto K, Ferrarini M, Uguccioni M, Piali L, Leone BE, Moser B, Rugarli C, Pardi R: Macrophages exposed to *Mycobacterium tuberculosis* release chemokines able to recruit selected leukocyte subpopulations: Focus on γ/δ cells. Immunology 2003;108:365–374.
130 Ladel CH, Hess J, Daugelat S, Mombaerts P, Tonegawa S, Kaufmann SH: Contribution of ααβ β and γ/δ T lymphocytes to immunity against *Mycobacterium bovis* bacillus Calmette Guerin: Studies with T cell receptor-deficient mutant mice. Eur J Immunol 1995;25:838–846.
131 Boom WH: γ/δ T cells and *Mycobacterium tuberculosis*. Microbes Infect 1999;1:187–195.
132 Cipriani B, Borsellino G, Poccia F, Placido R, Tramonti D, Bach S, Battistini L, Brosnan CF: Activation of C-C β-chemokines in human peripheral blood γ/δ T cells by isopentenyl pyrophosphate and regulation by cytokines. Blood 2000;95:39–47.
133 Murzenok PP, Matusevicius D, Freedman MS: γ/δ T cells in multiple sclerosis: Chemokine and chemokine receptor expression. Clin Immunol 2002;103:309–316.
134 Sallusto F, Kremmer E, Palermo B, Hoy A, Ponath P, Qin S, Forster R, Lipp M, Lanzavecchia A: Switch in chemokine receptor expression upon TCR stimulation reveals novel homing potential for recently activated T cells. Eur J Immunol 1999;29:2037–2045.
135 Brandes M, Willimann K, Lang AB, Nam KH, Jin C, Brenner MB, Morita CT, Moser B: Flexible migration program regulates γ/δ T-cell involvement in humoral immunity. Blood 2003;102: 3693–3701.
136 Dieli F, Poccia F, Lipp M, Sireci G, Caccamo N, Di Sano C, Salerno A: Differentiation of effector/memory Vδ2 T cells and migratory routes in lymph nodes or inflammatory sites. J Exp Med 2003;198:391–397.
137 Mackay CR: Dual personality of memory T cells. Nature 1999;401:659–660.
138 Sallusto F, Lenig D, Forster R, Lipp M, Lanzavecchia A: Two subsets of memory T lymphocytes with distinct homing potentials and effector functions. Nature 1999;401:708–712.
139 Eberl M, Engel R, Beck E, Jomaa H: Differentiation of human γ/δ T cells towards distinct memory phenotypes. Cell Immunol 2002;218:1–6.
140 Luster AD: Chemokines: Chemotactic cytokines that mediate inflammation. N Engl J Med 1998;338:436–445.
141 Colantonio L, Recalde H, Sinigaglia F, D'Ambrosio D: Modulation of chemokine receptor expression and chemotactic responsiveness during differentiation of human naive T cells into Th1 or Th2 cells. Eur J Immunol 2002;32:1264–1273.
142 Sallusto F, Lenig D, Mackay CR, Lanzavecchia A: Flexible programs of chemokine receptor expression on human polarized T helper 1 and 2 lymphocytes. J Exp Med 1998;187:875–883.
143 Wen L, Barber DF, Pao W, Wong FS, Owen MJ, Hayday A: Primary γ/δ cell clones can be defined phenotypically and functionally as Th1/Th2 cells and illustrate the association of CD4 with Th2 differentiation. J Immunol 1998;160:1965–1974.
144 Wesch D, Glatzel A, Kabelitz D: Differentiation of resting human peripheral blood γ/δ T cells toward Th1- or Th2-phenotype. Cell Immunol 2001;212:110–117.
145 Dagna L, Iellem A, Biswas P, Resta D, Tantardini F, Fortis C, Sabbadini MG, D'Ambrosio D, Manfredi AA, Ferrarini M: Skewing of cytotoxic activity and chemokine production, but not of chemokine receptor expression, in human type-1/-2 γ/δ T lymphocytes. Eur J Immunol 2002;32:2934–2943.
146 Wagner N, Lohler J, Kunkel EJ, Ley K, Leung E, Krissansen G, Rajewsky K, Muller W: Critical role for lβ7 integrins in formation of the gut-associated lymphoid tissue. Nature 1996;382: 366–370.
147 Uehara S, Grinberg A, Farber JM, Love PE: A role for CCR9 in T lymphocyte development and migration. J Immunol 2002;168:2811–2819.

148 Wurbel MA, Philippe JM, Nguyen C, Victorero G, Freeman T, Wooding P, Miazek A, Mattei MG, Malissen M, Jordan BR, Malissen B, Carrier A, Naquet P: The chemokine TECK is expressed by thymic and intestinal epithelial cells and attracts double- and single-positive thymocytes expressing the TECK receptor CCR9. Eur J Immunol 2000;30:262–271.
149 Zabel BA, Agace WW, Campbell JJ, Heath HM, Parent D, Roberts AI, Ebert EC, Kassam N, Qin SX, Zovko M, Larosa GJ, Yang LL, Soler D, Butcher EC, Ponath PD, Parker CM, Andrew DP: Human G protein-coupled receptor GPR-9-G/CC chemokine receptor 9 is selectively expressed on intestinal homing T lymphocytes, mucosal lymphocytes, and thymocytes and is required for thymus-expressed chemokine-mediated chemotaxis. J Exp Med 1999;190:1241–1255.
150 Kunkel EJ, Campbell JJ, Haraldsen G, Pan JL, Boisvert J, Roberts AI, Ebert EC, Vierra MA, Goodman SB, Genovese MC, Wardlaw AJ, Greenberg HB, Parker CM, Butcher EC, Andrew DP, Agace WW: Lymphocyte CC chemokine receptor 9 and epithelial thymus-expressed chemokine (TECK) expression distinguish the small intestinal immune compartment: Epithelial expression of tissue- specific chemokines as an organizing principle in regional immunity. J Exp Med 2000;192:761–767.
151 Papadakis KA, Prehn J, Nelson V, Cheng L, Binder SW, Ponath PD, Andrew DP, Targan SR: The role of thymus-expressed chemokine and its receptor CCR9 on lymphocytes in the regional specialization of the mucosal immune system. J Immunol 2000;165:5069–5076.
152 Wurbel MA, Malissen M, Guy-Grand D, Meffre E, Nussenzweig MC, Richelme M, Carrier A, Malissen B: Mice lacking the CCR9 CC-chemokine receptor show a mild impairment of early T- and B-cell development and a reduction in T-cell receptor γ/δ (+) gut intraepithelial lymphocytes. Blood 2001;98:2626–2632.
153 Uehara S, Song KM, Farber JM, Love PE: Characterization of CCR9 expression and CCL25/thymus-expressed chemokine responsiveness during T cell development: CD3(high) CD69(+) thymocytes and γ/δ TCR+ thymocytes preferentially respond to CCL25. J Immunol 2002;168: 134–142.
154 Marsal J, Svensson M, Ericsson A, Iranpour AH, Carramolino L, Marquez G, Agace WW: Involvement of CCL25 (TECK) in the generation of the murine small-intestinal CD8α α+CD3+ intraepithelial lymphocyte compartment. Eur J Immunol 2002;32:3488–3497.
155 Onai N, Kitabatake M, Zhang YY, Ishikawa H, Ishikawa S, Matsushima K: Pivotal role of CCL25 (TECK)-CCR9 in the formation of gut cryptopatches and consequent appearance of intestinal intraepithelial T lymphocytes. Int Immunol 2002;14:687–694.
156 Mora JR, Bono MR, Manjunath N, Weninger W, Cavanagh LL, Rosenblatt M, von Andrian UH: Selective imprinting of gut-homing T cells by Peyer's patch dendritic cells. Nature 2003;424: 88–93.
157 Kunkel EJ, Campbell DJ, Butcher EC: Chemokines in lymphocyte trafficking and intestinal immunity. Microcirculation 2003;10:313–323.
158 Cyster JG: Homing of antibody secreting cells. Immunol Rev 2003;194:48–60.
159 Nakayama T, Hieshima K, Izawa D, Tatsumi Y, Kanamaru A, Yoshie O: Cutting edge: Profile of chemokine receptor expression on human plasma cells accounts for their efficient recruitment to target tissues. J Immunol 2003;170:1136–1140.
160 Kunkel EJ, Kim CH, Lazarus NH, Vierra MA, Soler D, Bowman EP, Butcher EC: CCR10 expression is a common feature of circulating and mucosal epithelial tissue IgA Ab-secreting cells. J Clin Invest 2003;111:1001–1010.
161 Soler D, Humphreys TL, Spinola SM, Campbell JJ: CCR4 versus CCR10 in human cutaneous TH lymphocyte trafficking. Blood 2003;101:1677–1682.
162 Campbell JJ, Haraldsen G, Pan J, Rottman J, Qin S, Ponath P, Andrew DP, Warnke R, Ruffing N, Kassam N, Wu L, Butcher EC: The chemokine receptor CCR4 in vascular recognition by cutaneous but not intestinal memory T cells. Nature 1999;400:776–780.
163 Hudak S, Hagen M, Liu Y, Catron D, Oldham E, Mcevoy LM, Bowman EP: Immune surveillance and effector functions of CCR10(+) skin homing T cells. J Immunol 2002;169:1189–1196.
164 Kapp JA, Kapp LM, McKenna KC: γ/δ T cells play an essential role in several forms of tolerance. Immunol Res 2004;29:93–102.
165 Kapp JA, Kapp LM, McKenna KC, Lake JP: γδ T-cell clones from intestinal intraepithelial lymphocytes inhibit development of CTL responses ex vivo. Immunology 2004;111:155–164.

166 McMenamin C, Pimm C, McKersey M, Holt PG: Regulation of IgE responses to inhaled antigen in mice by antigen-specific γδ T cells. Science 1994;265:1869–1871.
167 Mincheva-Nilsson L, Kling M, Hammarstrom S, Nagaeva O, Sundqvist KG, Hammarstrom ML, Baranov V: γδ T cells of human early pregnancy decidua: Evidence for local proliferation, phenotypic heterogeneity, and extrathymic differentiation. J Immunol 1997;159:3266–3277.
168 Barakonyi A, Kovacs KT, Miko E, Szereday L, Varga P, Szekeres-Bartho J: Recognition of nonclassical HLA class I antigens by γ/δ T cells during pregnancy. J Immunol 2002;168: 2683–2688.
169 Nagaeva O, Jonsson L, Mincheva-Nilsson L: Dominant IL-10 and TGF-β mRNA expression in γ/δ T cells of human early pregnancy decidua suggests immunoregulatory potential. Am J Reprod Immunol 2002;48:9–17.
170 Holt PG, Sly PD: γ/δ T cells provide a breath of fresh air for asthma research. Nat Med 1999;5:1127–1128.
171 McMenamin C, McKersey M, Kuhnlein P, Hunig T, Holt PG: γδ T cells down-regulate primary IgE responses in rats to inhaled soluble protein antigens. J Immunol 1995;154:4390–4394.
172 Lahn M, Kanehio A, Takeda K, Joetham A, Schwarze J, Kohler G, O'Brien R, Gelfand EW, Born W: Negative regulation of airway responsiveness that is dependent on γ/δ T cells and independent of ααβ T cells. Nat Med 1999;5:1150–1156.
173 Lahn M: The role of γ/δ T cells in the airways. J Mol Med 2000;78:409–425.
174 Workalemahu G, Foerster M, Kroegel C: Expression and synthesis of fibroblast growth factor-9 in human γδ T-lymphocytes. Response to isopentenyl pyrophosphate and TGF-β1/IL-15. J Leukoc Biol 2004;75:657–663.
175 Boismenu R, Havran WL: Modulation of epithelial cell growth by intraepithelial γ/δ T cells. Science 1994;266:1253–1255.
176 Havran WL: A role for epithelial γ/δ T cells in tissue repair. Immunol Res 2000;21:63–69.
177 Chen YP, Chou K, Fuchs E, Havran WL, Boismenu R: Protection of the intestinal mucosa by intraepithelial γ/δ T cells. Proc Natl Acad Sci USA 2002;99:14338–14343.
178 Workalemahu G, Foerster M, Kroegel C, Braun RK: Human γ/δ-T lymphocytes express and synthesize connective tissue growth factor: Effect of IL-15 and TGF-β1 and comparison with ααβ-T lymphocytes. J Immunol 2003;170:153–157.
179 Yang H, Antony PA, Wildhaber BE, Teitelbaum DH: Intestinal intraepithelial lymphocyte gammadelta-T cell-derived keratinocyte growth factor modulates epithelial growth in the mouse. J Immunol 2004;172:4151–4158.
180 Hoffmann JC, Peters K, Henschke S, Herrmann B, Pfister K, Westermann J, Zeitz M: Role of T lymphocytes in rat 2,4,6-trinitrobenzene sulphonic acid (TNBS) induced colitis: Increased mortality after γ/δ T cell depletion and no effect of ααββT cell depletion. Gut 2001;48: 489–495.
181 Guan H, Zu G, Slater M, Elmets C, Xu H: γ/δ T cells regulate the development of hapten-specific CD8+ effector T cells in contact hypersensitivity responses. J Invest Dermatol 2002;119: 137–142.
182 Girardi M, Lewis J, Glusac E, Filler RB, Geng L, Hayday AC, Tigelaar RE: Resident skin-specific γ/δ T cells provide local, nonredunda nt regulation of cutaneous inflammation. J Exp Med 2002;195:855–867.
183 Skeen MJ, Rix EP, Freeman MM, Ziegler HK: Exaggerated proinflammatory and Th1 responses in the absence of γ/δ T cells after infection with *Listeria monocytogenes*. Infect Immun 2001;69:7213–7223.
184 Egan PJ, Carding SR: Downmodulation of the inflammatory response to bacterial infection by γ/δ T cells cytotoxic for activated macrophages. J Exp Med 2000;191:2145–2158.
185 Seo N, Egawa K: Suppression of cytotoxic T lymphocyte activity by γ/δ T cells in tumor-bearing mice. Cancer Immunol Immunother 1995;40:358–366.
186 Seo N, Tokura Y, Takigawa M, Egawa K: Depletion of IL-10- and TGF-β-producing regulatory γ/δ T cells by administering a daunomycin-conjugated specific monoclonal antibody in early tumor lesions augments the activity of CTLs and NK cells. J Immunol 1999;163:242–249.
187 Ke Y, Kapp LM, Kapp JA: Inhibition of tumor rejection by γδ T cells and IL-10. Cell Immunol 2003;221:107–114.

188 Uezu K, Kawakami K, Miyagi K, Kinjo Y, Kinjo T, Ishikawa H, Saito A: Accumulation of γ/δ T cells in the lungs and their regulatory roles in Th1 response and host defense against pulmonary infection with *Cryptococcus neoformans*. J Immunol 2004;172:7629–7634.
189 Kopp E, Medzhitov R: Recognition of microbial infection by Toll-like receptors. Curr Opin Immunol 2003;15:396–401.
190 Caramalho I, Lopes-Carvalho T, Ostler D, Zelenay S, Haury M, Demengeot J: Regulatory T cells selectively express toll-like receptors and are activated by lipopolysaccharide. J Exp Med 2003;197:403–411.
191 Mokuno Y, Matsuguchi T, Takano M, Nishimura H, Washizu J, Ogawa T, Takeuchi O, Akira S, Nimura Y, Yoshikai Y: Expression of toll-like receptor 2 on γ/δ T cells bearing invariant Vγ6/Vδ1 induced by *Escherichia coli* infection in mice. J Immunol 2000;165:931–940.
192 Alexopoulou L, Holt AC, Medzhitov R, Flavell RA: Recognition of double-stranded RNA and activation of NF-kappaB by Toll-like receptor 3. Nature 2001;413:732–738.
193 Kunzmann V, Kretzschmar E, Herrmann T, Wilhelm M: Polyinosinic-polycytidylic acid-mediated stimulation of human γδ T cells via CD11c dendritic cell-derived type I interferons. Immunology 2004;112:369–377.

Wolfgang Holtmeier, MD
Medizinische Klinik I, Johann Wolfgang Goethe-Universität
Theodor-Stern-Kai 7, DE–60590 Frankfurt/Main (Germany)
Tel. +49 69 63016218, Fax +49 69 63016448, E-Mail W.Holtmeier@em.uni-frankfurt.de

Author Index

Subject Index